T0213980

3D Printing in Oral & Maxillofacial Surgery

Lobat Tayebi • Reza Masaeli
Kavosh Zandsalimi

3D Printing in Oral & Maxillofacial Surgery

 Springer

Lobat Tayebi
School of Dentistry
Marquette University
Milwaukee, WI
USA

Reza Masaeli
School of Dentistry
Tehran University of Medical Sciences
Tehran
Iran

Kavosh Zandsalimi
Life Sciences Engineering
University of Tehran
Tehran
Iran

ISBN 978-3-030-77789-0   ISBN 978-3-030-77787-6   (eBook)
https://doi.org/10.1007/978-3-030-77787-6

This Springer imprint is published by the registered company Springer Nature Switzerland AG
The registered company address is: Gewerbestrasse 11, 6330 Cham, Switzerland

Acknowledgments

Part of the research reported in this paper was supported by the National Institute of Dental and Craniofacial Research of the National Institutes of Health under award number R15DE027533.

The content is solely the responsibility of the authors and does not necessarily represent the official views of the National Institutes of Health.

The authors would like to thank Professor Arndt Guentsch for his valuable discussion, inputs, and write-up on periodontal tissue regeneration presented in Chap. 7, and Mr. Mohammad Sabzevarian for creating illustrations presented in Chaps. 2 and 3.

Contents

A future in medicine is envisioned where medical treatment will become highly personalized with treatment modalities that are patient-specific, rather than a "one size fits all" approach. Such a future requires applications that possess quick fabrication of constructs with complex shapes and high performance. This functionality will necessitate machinery with increased sensitivity, specificity, tunability, and that is customizable. While this technology has yet to be fully realized, recent advances in three-dimensional (3D) printing may enable on-demand and patient-specific medical treatments, particularly in the field of oral and maxillofacial surgery. Of course, interdisciplinary teams of clinicians, engineers, and scientists must work collaboratively to realize such a vision to repair, regenerate, and reconstruct dental, oral, and craniofacial tissues.

The reconstruction of the oromaxillofacial complex is a difficult task. The presence of different tissue types within a small region is one of the special complexities of the oral and maxillofacial region. This proximity often results in tissue loss that requires the reconstruction of various tissues to manage developmental, pathological, and traumatic defects [1–5].

For example, conditions such as oral cancer, locally aggressive benign tumors (such as ameloblastoma), jaw cysts, osteomyelitis, gunshot wounds, osteoradionecrosis, and medication-related osteonecrosis of the jaw (MRONJ) require extensive surgical resection and reconstruction of the affected area, in addition to the removal of tissues from surrounding healthy bone and soft tissues. Smaller defects, such as those seen in dentoalveolar ridges after trauma, periodontal disease, and partial anodontia cases also require challenging surgical reconstruction methods.

Currently, reconstructive methods involve dentoalveolar ridges, tissue expanders, autogenous bone grafting, allogenic and xenogenic bone grafting, application of growth factors (platelet-rich or fibrin-rich plasma, or bone morphogenic proteins), skin grafting, local and regional tissue flaps, and microvascular free flaps [3, 6–24]. The specific procedure selected for a craniofacial deformity depends on the patient's factors (including health status, history of radiation therapy, smoking

© Springer Nature Switzerland AG 2021
L. Tayebi et al., *3D Printing in Oral & Maxillofacial Surgery*,
https://doi.org/10.1007/978-3-030-77787-6_1

status, etc.), as well as the characteristics of the defect including its size, the type of tissues involved, and location [25].

Complex craniofacial defects that remain after the treatment of pathological conditions differ in the number of tissues that are involved. Some defects involve the bone only, while others have missing soft tissue as well. This soft tissue could represent an area of the oral cavity or the skin of the face. The complexity of the soft-tissue defect may require a significant modification in the therapy. If the defect cannot be closed using local measures with the advancement of soft tissue flaps, the area will need to be closed using a local or regional flap (pectoralis major, temporalis, or others), microvascular flap (radial forearm free flap, fibula free flap, or others), or using a skin graft followed by bony reconstruction at a later stage. Tissue expanders can be used for defects that have been closed initially but will likely be under significant tension when a bone graft is placed [23, 26–32].

The defect size is a critical factor in determining the type of reconstructive technique selected. Defects larger than 2.5–5 cm in the mandible, for example, typically require a microvascular free flap reconstruction. The size of the defect hinders the vascularity of non-vascularized bone graft (from the anterior or posterior iliac crest as an example), and the graft will most likely undergo necrosis, followed by post-surgical infection [33]. Distraction osteogenesis is indicated in instances where there is a sufficient stock of bone that can be distracted into the position that is needed. The advantage of this technique is that it not only generates bone, but also generates soft tissue. It requires a significant amount of time and numerous surgical interventions. The device is also cumbersome, and often esthetically unpleasant, however, repetitive surgeries are very impactful on patients [24, 34, 35].

Microvascular-free flaps revolutionized the reconstruction of craniofacial deformities. Post-ablative tumor or trauma surgeries create disfiguring defects that leave the affected patients unable to return to daily functions, including basic speech, swallowing, and facial animation [36]. Microvascular surgery has undergone significant improvement over the past 50 years [37, 38]. The length of the surgical procedure has decreased significantly, allowing this procedure to be performed in patients with less-than-ideal health conditions. The recovery periods have also been reduced, and patients can return to their active daily lives soon after their surgical procedures. The procedure, however, remains high-risk. The length of the procedure associated swelling, and the need for tracheotomies as well as percutaneous endoscopic gastrostomy tube (PEG) tubes makes it troublesome for many patients and their family members. The length of hospital stay increases the risks of hospital-acquired infections, and tracheotomies themselves have their risks [39].

3D printing is currently being intensely studied for use in a diverse set of applications, including the production of functional biomedical materials and devices for dental and orthopedic applications.

Applying 3D printing technology to repair, reconstruct, and regenerate oral and maxillofacial defects can help develop unique solutions that address the issues of cost and customization through designs that are patient-specific and conformable to the patients' unique needs. The primary focus can be directed at creating novel bioactive scaffolds for in situ tissue repair and generation. These scaffolds can be tailored to meet the patient's existing condition, age, and prior medical history, and are designed for tension-free soft tissue closure, enhanced vascularity and tissue formation, and the prevention of micromobility.

In terms of printing cells, the bioink is essentially a cell-laden hydrogel of the appropriate viscosity capable of being extruded under pressure through a microscale nozzle orifice or a micro-needle at temperatures around 37 °C to maintain cell viability [40]. The rheological properties of the biomaterial, extrusion temperature, nozzle type used, and applied pressure are the critical parameters that affect the physical and biological characteristics of the printed construct [41–43].

Bioinks are an integral part of bioprinting technology [40, 44–48]. It is not the bioprinting process parameters alone, but the material-process interactions that govern the viability and success of the resultant constructs. Hence, developing appropriate bioinks and comprehensively characterizing their rheological, mechanical, and biological characteristics are critical to the success of bioprinting. It is accepted that this development and characterization will have to be cell/tissue and process specific [40, 49].

Cells can be conventionally grown in monocultures or co-cultures and can be seeded on or encapsulated inside a biocompatible degradable scaffold. The main idea is that the degradation of these materials by the cells over time and the replacement of them with the extracellular matrix (ECM) will form living tissues. Since the biomaterial structure and other microenvironmental cues can affect the functional behavior of this in vitro tissue, the materials must be chosen carefully [50–52].

Microfluidic-based technologies present interesting opportunities for tuning the cell-materials interactions [52]. In the content of cell encapsulation, one may use the microfluidic technique for cell encapsulation in natural or synthetic materials. For example, encapsulating stem cells in gel-based microdroplets can be used to form cell-containing fibers and microtubes from gel precursor solutions [53–56]. These cell-based constructs can be used in the bioprinting process to make functional tissue constructs for oral and maxillofacial defect reconstruction. Such microfluidic approaches can offer the ability to continuously alter the particle shape and size, simply by changing the flow rate of the prepolymer solution, and does so reproducibly without the need for masks and masters [57].

Thus, one can generate small and monodisperse growth factor-loading nanoparticles by creating a narrowly defined mixing regime [58–63]. This well-controlled mixing regime can be precisely adjusted with the microfluidic platform (primarily through flow ratio and velocity) and is crucial in creating monodisperse nanoparticles with tunable properties, including drug encapsulation efficiency and release rate [60, 61, 63].

These cell-loaded microparticles, as well as growth-factor-loaded nanoparticles, can be incorporated inside a scaffold during or after the printing process. The call and growth factor selection must be based on the specific needs for different conditions of the oral and maxillofacial defects.

We anticipate that 3D printing, in combination with the cell-loaded and growth-factor loaded micro/nanoparticles, as described above, can be helpful for the following clinical problems that require difficult clinical decisions, are lengthy surgical procedures, pricey in terms of medical costs, and have less than the ideal microvascular free flaps.

1. Mandibular bony defects >5 cm (with intact soft tissues).
2. Mandibular and maxillary defects including a large amount of soft tissue loss.
3. Maxillary and mandibular ridge resorption due to periodontal disease/trauma/congenital defects.
4. Anterior mandibular (chin) defects.
5. Other critical defects in the esthetic regions of the oral cavity (anterior maxilla).

Different designs and technical approaches of 3D printing will be needed for different conditions as well as the appropriate use of cells, growth factors, and biomaterials such as polymers, ceramics, natural and synthetic bioplastics, proteins and biomolecules, living cells, and growth factors, as well as their hybrid structures.

Fabricating FDA approved, vascularized, customized 3D scaffolds that are infused with relevant stem cells and growth factors to replace oral and maxillofacial defects can significantly improve the quality of the life of patients with the above clinical problems, who are affected by numerous dental and craniofacial pathologic conditions (such as cancer, benign locally aggressive tumors, medication-related osteonecrosis, radiation-related osteonecrosis, congenital/acquired loss of alveolar bone and/or teeth, and avulsive trauma). Additionally, healthcare costs will be reduced, the quality of life will be improved, and posttreatment complications will decrease. The process will be a procedure with lower risk allowing the surgeons to provide ideal services to patients with increased health risks (such as those with kidney disease, cardiovascular disease, and coagulopathies).

This is of course an ambitious goal! We recognize this and embrace the challenge. To go in this direction, one must be practical and aware of the constraints and obstacles. Linking practicing clinicians with accomplished and established investigators from different fields with unique abilities, skill sets, and specialties is necessary to have progress in such interdisciplinary translational research to achieve the goal of using 3D printing in oral and maxillofacial surgery. More specifically, investigators in the field of biomaterials, biomedicine, drug delivery, regenerative medicine, tissue engineering, and organ development must closely work with oral and maxillofacial surgeons. Basically, as we will see in this book, many scientists have individually significantly advanced the field. However, their developed methods and technologies have never been applied or directed towards the unified goal of repairing and regenerating oral and maxillofacial tissues in clinics. For example, effective methods exist for the 3D printing of various constructs and scaffolds, loading growth factors into scaffolds, or the directed vascularization of synthetic tissues. But they have not been applied together to advance the healing of oral and maxillofacial defects. Owing to the interdisciplinary nature of this goal, an infrastructure is required to facilitate and direct the active collaboration of clinicians, engineers, and scientists from various disciplines in a highly cooperative, entrepreneurial, interactive, and creative environment.

This book will begin by introducing the various 3D printing methods that can be applicable in oral and maxillofacial surgery including Stereolithography (SLA), selective laser sintering (SLS), fused deposition modeling (FDM), photopolymer jetting (PPJ), and powder binder printing (PBP). For each method, after explaining the process and investigating its advantages and limitations, their oral and maxillofacial applications are addressed. Printable materials in four groups of metals, ceramics, synthetic polymers, and hydrogels are elaborated, and their potential applications in oral and maxillofacial surgeries are discussed.

3D bioprinting is the subject of the third chapter, in which after detailed consideration of different types of 3D bioprinting and their oromaxillofacial applications, this technique is concluded to be impractical for oral and maxillofacial surgery. This conclusion is relevant to the limitation of bioprinting itself and also conventional clinical settings. To make bioprinting practical for oral and maxillofacial surgery, translation of the fabricating 3D bioprinted tissue constructs from the laboratory to the clinic is necessary.

The next chapter is dedicated to the application of 3D printed medical models for oral and maxillofacial surgery. To fabricate such models, the SLA method is more useful than other 3D printing techniques. Dental, TMJ, mandibular, facial, and skull 3D printing models are discussed in this chapter.

The fifth chapter argues the application of 3D printing in the manufacturing of dental instruments including surgical guides, splints, oral and maxillofacial implants, dental implants, total jaw prostheses, crowns, and dentures. The tissue engineering approach in oral and maxillofacial reconstruction is described in Chap. 6. Various examples of hard and soft tissue printing and bioprinting in oromaxillofacial surgery, as well as their requirements and advances, are explained in this chapter. In conventional bone tissue engineering, a biodegradable and porous bone scaffold is implanted in the injured area, enabling bone cells to grow inside the porous materials of the scaffolds over time. Employing a biodegradable bone tissue engineering scaffold has many advantages such as facilitating the regeneration of the bone cells by different mechanisms. However, despite the advantages, commercially, metal implants are still preferred for critical-sized oral and maxillofacial defects. This is due to the lack of mechanical strength of the current bone scaffolds as they are not able to bear the load of the body. This chapter addresses this problem by introducing a hybrid reinforced scaffold that can significantly enhance the mechanical strength of the current scaffolds.

The design is inspired by reinforced concrete, in which reinforcing bars (rebars) are embedded passively in the concrete; thus, the concrete's relatively low strength is counteracted by the inclusion of the reinforcement. Our scaffold is composed of two components: (1) Skeleton: nonporous and slowly biodegradable constituent undertaking mechanical necessities of the scaffolds, and (2) Host Component: porous and biodegradable constituent undertaking biological necessities of the scaffolds. The 3D printed Skeleton is placed into the Host Component to reinforce it.

The last chapter of this book focuses on oral and maxillofacial multi- and interfacial tissue defects and describes how 3D printing can be helpful for the regeneration of such defects due to its nature of layer-by-layer material deposition. The regeneration of osteochondral, osteomucosal, and periodontal complexes are discussed separately. Additionally, the fabrication and use of 3D printed Guided Bone Regeneration (GBR) and Guided Tissue Engineering (GTR) membranes in the reconstruction of multi and interfacial -tissue constructs in the oral and maxillofacial region are elaborated. An important challenge in the reconstruction of multi- and interfacial tissue defects is related to the adhesion of soft and hard tissues. An adhesion process is explained in this chapter based on an in situ biocompatible chemical process, which can be very helpful in the reconstruction of oral and maxillofacial multi- and interfacial tissue defects.

In summary, there are valuable advances in using 3D printing techniques for oral and maxillofacial surgery. However, major challenges are needed to be resolved before being able to widely and practically apply this approach in clinics.

References

1. Jalbert F, Boetto S, Nadon F, Lauwers F, Schmidt E, Lopez R. One-step primary reconstruction for complex craniofacial resection with PEEK custom-made implants. J Cranio-Maxillofac Surg. 2014;42(2):141–8.
2. Bagheri SC, Bell B, Khan HA. Current therapy in oral and maxillofacial surgery. Elsevier; 2011.
3. Herford AS, Tandon R, Stevens TW, Stoffella E, Cicciu M. Immediate distraction osteogenesis: the sandwich technique in combination with rhBMP-2 for anterior maxillary and mandibular defects. J Craniofac Surg. 2013;24(4):1383–7.
4. Chopra K, Folstein MK, Manson PN, Gastman BR. Complex craniofacial reconstruction using stereolithographic modeling. Ann Plast Surg. 2014;72(1):59–63.
5. Rana M, Warraich R, Kokemüller H, Lemound J, Essig H, Tavassol F, et al. Reconstruction of mandibular defects—clinical retrospective research over a 10-year period. Head Neck Oncol. 2011;3:23.
6. O'Connell DA, Teng MS, Mendez E, Futran ND. Microvascular free tissue transfer in the reconstruction of scalp and lateral temporal bone defects. Craniomaxillofac Trauma Reconstr. 2011;4(4):179.
7. Jensen OT, Leopardi A, Gallegos L. The case for bone graft reconstruction including sinus grafting and distraction osteogenesis for the atrophic edentulous maxilla. J Oral Maxillofac Surg. 2004;62(11):1423–8.
8. Polley JW, Figueroa AA. Distraction osteogenesis: its application in severe mandibular deformities in hemifacial microsomia. J Craniofac Surg. 1997;8(5):422–30.
9. Herford AS, Boyne PJ. Reconstruction of mandibular continuity defects with bone morphogenetic protein-2 (rhBMP-2). J Oral Maxillofac Surg. 2008;66(4):616–24.
10. Carter TG, Brar PS, Tolas A, Beirne OR. Off-label use of recombinant human bone morphogenetic protein-2 (rhBMP-2) for reconstruction of mandibular bone defects in humans. J Oral Maxillofac Surg. 2008;66(7):1417–25.
11. Alden TD, Beres EJ, Laurent JS, Engh JA, Das S, London SD, et al. The use of bone morphogenetic protein gene therapy in craniofacial bone repair. J Craniofac Surg. 2000;11(1):24–30.
12. Moghadam HG, Urist MR, Sandor GK, Clokie CM. Successful mandibular reconstruction using a BMP bioimplant. J Craniofac Surg. 2001;12(2):119–27.
13. Shields LB, Raque GH, Glassman SD, Campbell M, Vitaz T, Harpring J, et al. Adverse effects associated with high-dose recombinant human bone morphogenetic protein-2 use in anterior cervical spine fusion. Spine. 2006;31(5):542–7.
14. Arosarena OA, Collins WL. Bone regeneration in the rat mandible with bone morphogenetic protein-2: a comparison of two carriers. Otolaryngol Head Neck Surg. 2005;132(4):592–7.

15. Feng Z, Zhao J, Zhou L, Dong Y, Zhao Y. Modified animal model and computer-assisted approach for dentoalveolar distraction osteogenesis to reconstruct unilateral maxillectomy defect. J Oral Maxillofac Surg. 2009;67(10):2266–74.
16. Feng Y, Fang B, Shen G, Xia Y, Lou X. Reconstruction of partial maxillary defect with intraoral distraction osteogenesis assisted by miniscrew implant anchorages. Oral Surg Oral Med Oral Pathol Oral Radiol Endod. 2010;110(3):e1–7.
17. Moiyadi AV, Ghazwan QA, Pai PS, Kelkar G, Nair D, Yadav PS. Free anterolateral thigh flap for reconstruction of complex craniofacial defects after resection of tumors of the fronto-orbitomaxillary complex. J Craniofac Surg. 2012;23(3):836–41.
18. Aksu AE, Dursun E, Calis M, Ersu B, Safak T, Tözüm TF. Intraoral use of extraoral implants for oral rehabilitation of a pediatric patient after resection of Ewing sarcoma of the mandible and reconstruction with iliac osteocutaneous free flap. J Craniofac Surg. 2014;25(3):930–3.
19. Heffelfinger R, Murchison AP, Parkes W, Krein H, Curry J, Evans JJ, et al. Microvascular free flap reconstruction of orbitocraniofacial defects*. Orbit. 2013;32(2):95–101.
20. Reyes C, Mason E, Solares CA. Panorama of reconstruction of skull base defects: from traditional open to endonasal endoscopic approaches, from free grafts to microvascular flaps. Int Arch Otorhinolaryngol. 2014;18:179–86.
21. Schusterman MA, Miller MJ, Reece GP, Kroll SS, Marchi M, Goepfert H. A single center's experience with 308 free flaps for repair of head and neck cancer defects. Plast Reconstr Surg. 1994;93(3):472–8.
22. Fisher J, Jackson IT. Microvascular surgery as an adjunct to craniomaxillofacial reconstruction. Br J Plast Surg. 1989;42(2):146–54.
23. Losken A, Carlson GW, Culbertson JH, Scott Hultman C, Kumar AV, Jones GE, et al. Omental free flap reconstruction in complex head and neck deformities. Head Neck. 2002;24(4):326–31.
24. Watanabe Y, Sasaki R, Ando T, Okano T, Akizuki T. Multistage bone-charged distraction osteogenesis for aesthetic reconstruction of an extensive bone deficiency in the mandible. J Craniofac Surg. 2012;23(1):175–7.
25. Hurvitz K, Kobayashi M, Evans G. Current options in head and neck reconstruction. Plast Reconstr Surg. 2006;118:122e–33e.
26. Kawashima T, Yamada A, Ueda K, Asato H, Harii K. Tissue expansion in facial reconstruction. Plast Reconstr Surg. 1994;94:944–50.
27. Radovan C. Tissue expansion in soft-tissue reconstruction. Plast Reconstr Surg. 1984;74(4):482–90.
28. Khouri RK, Ozbek MR, Hruza GJ, Young VL. Facial reconstruction with prefabricated induced expanded (PIE) supraclavicular skin flaps. Plast Reconstr Surg. 1995;95(6):1007–15.
29. Gruss JS, Antonyshyn O, Phillips JH. Early definitive bone and soft-tissue reconstruction of major gunshot wounds of the face. Plast Reconstr Surg. 1991;87(3):436–50.
30. Bakhshaeekia A. Ten-year experience in face and neck unit reconstruction using tissue expanders. Burns. 2013;39(3):522–7.
31. Kheradmand AA, Garajei A, Motamedi MHK. Nasal reconstruction: experience using tissue expansion and forehead flap. J Oral Maxillofac Surg. 2011;69(5):1478–84.
32. Hofer SO, Mureau MA. Improving outcomes in aesthetic facial reconstruction. Perioper Nurs Clin. 2011;6(2):147–58.
33. Chim H, Salgado CJ, Mardini S, Chen H-C. Reconstruction of mandibular defects. Semin Plast Surg. 2010;24:188–97.
34. Vega LG, Bilbao A. Alveolar distraction osteogenesis for dental implant preparation: an update. Oral Maxillofac Surg Clin North Am. 2010;22:369–85.
35. Aizenbud D, Hazan-Molina H, Cohen M, Rachmiel A. Combined orthodontic temporary Anchorage devices and surgical Management of the Alveolar Ridge Augmentation Using Distraction Osteogenesis. J Oral Maxillofac Surg. 2012;70(8):1815–26.
36. Hidalgo DA. Fibular free flap: a new method of mandible reconstruction. Plast Reconstr Surg. 1989;84:71–9.
37. Khouri RK. Free flap surgery. The second decade. Clin Plast Surg. 1992;19:757–61.
38. Buncke HJ, Chang DW. History of microsurgery. Semin Plast Surg. 2003:5–15.

39. Durbin CG Jr. Early complications of tracheostomy. Respir Care. 2005;50:511–5.
40. Pataky K, Braschler T, Negro A, Renaud P, Lutolf MP, Brugger J. Microdrop printing of hydrogel bioinks into 3D tissue-like geometries. Adv Mater. 2012;24(3):391–6.
41. Seol Y-J, Kang H-W, Lee SJ, Atala A, Yoo JJ. Bioprinting technology and its applications. Eur J Cardiothorac Surg. 2014;46:342–8.
42. Xu T, Zhao W, Atala A, Yoo JJ, editors. Bio-printing of living organized tissues using an inkjet technology. In: NIP & digital fabrication conference. Society for Imaging Science and Technology; 2006.
43. Xu T, Zhao W, Zhu J-M, Albanna MZ, Yoo JJ, Atala A. Complex heterogeneous tissue constructs containing multiple cell types prepared by inkjet printing technology. Biomaterials. 2013;34(1):130–9.
44. Levato R, Visser J, Planell JA, Engel E, Malda J, Mateos-Timoneda MA. Biofabrication of tissue constructs by 3D bioprinting of cell-laden microcarriers. Biofabrication. 2014;6(3):035020.
45. Mironov V, Kasyanov V, Drake C, Markwald RR. Organ printing: promises and challenges. Regen Med. 2008;3:93–103.
46. Campbell PG, Weiss LE. Tissue engineering with the aid of inkjet printer. Expert Opin Biol Ther. 2007;7(8):1123–7.
47. Mironov V, Visconti RP, Kasyanov V, Forgacs G, Drake CJ, Markwald RR. Organ printing: tissue spheroids as building blocks. Biomaterials. 2009;30(12):2164–74.
48. Mironov V, Kasyanov V, Markwald RR. Organ printing: from bioprinter to organ biofabrication line. Curr Opin Biotechnol. 2011;22(5):667–73.
49. Melchels FP, Domingos MA, Klein TJ, Malda J, Bartolo PJ, Hutmacher DW. Additive manufacturing of tissues and organs. Prog Polym Sci. 2012;37(8):1079–104.
50. Chung BG, Kang L, Khademhosseini A. Micro-and nanoscale technologies for tissue engineering and drug discovery applications. Expert Opin Drug Discov. 2007;2:1653–68.
51. Edalat F, Bae H, Manoucheri S, Cha JM, Khademhosseini A. Engineering approaches toward deconstructing and controlling the stem cell environment. Ann Biomed Eng. 2012;40(6):1301–15.
52. Gauvin R, Khademhosseini A. Microscale technologies and modular approaches for tissue engineering: moving toward the fabrication of complex functional structures. ACS Nano. 2011;5(6):4258–64.
53. Hwang C, Park Y, Park J, Lee K, Sun K, Khademhosseini A, et al. Controlled cellular orientation on PLGA microfibers with defined diameters. Biomed Microdevices. 2009;11(4):739–46.
54. Kang E, Jeong GS, Choi YY, Lee KH, Khademhosseini A, Lee S-H. Digitally tunable physicochemical coding of material composition and topography in continuous microfibres. Nat Mater. 2011;10(11):877–83.
55. Ling Y, Rubin J, Deng Y, Huang C, Demirci U, Karp JM, et al. A cell-laden microfluidic hydrogel. Lab Chip. 2007;7(6):756–62.
56. Nichol JW, Khademhosseini A. Modular tissue engineering: engineering biological tissues from the bottom up. Soft Matter. 2009;5(7):1312–9.
57. Chung BG, Lee K-H, Khademhosseini A, Lee S-H. Microfluidic fabrication of microengineered hydrogels and their application in tissue engineering. Lab Chip. 2012;12(1):45–59.
58. Hasani-Sadrabadi MM, Majedi FS, VanDersarl JJ, Dashtimoghadam E, Ghaffarian SR, Bertsch A, et al. Morphological tuning of polymeric nanoparticles via microfluidic platform for fuel cell applications. J Am Chem Soc. 2012;134(46):18904–7.
59. Hasani-Sadrabadi MM, VanDersarl JJ, Dashtimoghadam E, Bahlakeh G, Majedi FS, Mokarram N, et al. A microfluidic approach to synthesizing high-performance microfibers with tunable anhydrous proton conductivity. Lab Chip. 2013;13(23):4549–53.
60. Hasani-Sadrabadi MM, Karimkhani V, Majedi FS, Van Dersarl JJ, Dashtimoghadam E, Afshar-Taromi F, et al. Microfluidic-assisted self-assembly of complex dendritic polyethylene drug delivery Nanocapsules. Adv Mater. 2014;26(19):3118–23.
61. Majedi FS, Hasani-Sadrabadi MM, Emami SH, Shokrgozar MA, VanDersarl JJ, Dashtimoghadam E, et al. Microfluidic assisted self-assembly of chitosan based nanoparticles as drug delivery agents. Lab Chip. 2013;13(2):204–7.

62. Majedi FS, Hasani-Sadrabadi MM, Emami SH, Taghipoor M, Dashtimoghadam E, Bertsch A, et al. Microfluidic synthesis of chitosan-based nanoparticles for fuel cell applications. Chem Commun. 2012;48(62):7744–6.
63. Majedi FS, Hasani-Sadrabadi MM, VanDersarl JJ, Mokarram N, Hojjati-Emami S, Dashtimoghadam E, et al. On-Chip fabrication of paclitaxel-loaded chitosan nanoparticles for Cancer therapeutics. Adv Funct Mater. 2014;24(4):432–41.

3D Printing Methods Applicable in Oral and Maxillofacial Surgery

2

2.1 Introduction

From the distant past to the present day, human beings have applied various techniques to fabricate objects. The most commonly used fabrication methods include casting, molding, forming, and machining, which follow a "subtractive" approach; in other words, a solid block of material is cut consecutively to construct the desired object. Over the centuries, subtractive methods have had outstanding advancements that improve the quality of final products and increase production efficiency [1].

Not so long ago, production was limited, as it relied on skilled labor. With the onset of the Industrial Revolution, quick yet high-quality production became a necessity. The invention of computer-aided design/manufacturing (CAD/CAM) systems in the 1960s increased the quality and efficiency of production lines and reduced workforce dependency. The automotive and aerospace industries were the pioneers in the automation of their manufacturing operations using CAD/CAM systems [2].

In the 1970s, the advent of 3D medical imaging techniques, that is, computed tomography (CT) and magnetic resonance imaging (MRI), paved the way for the usage of CAD/CAM systems in medicine and dentistry. The emergent imaging methods provided a robust way to obtain invaluable information regarding the internal structure of tissues and organs [3, 4]. The combination of new imaging methods and CAD/CAM systems made it possible to construct tissue substitutes based on the acquired 3D anatomical data.

The first attempts to fabricate functional dental restorations via CAD/CAM began by Duret et al. in 1971 [5]. In the late 1980s, the first chairside CAD/CAM machines were utilized within dental offices and laboratories. The main benefit of this technology over the conventional methods in restorative dentistry was the reduction in chair time by eliminating the need for impressions. However, the high cost of the equipment and restrictions on fabricating the prostheses with fine anatomical details prevented the pervasiveness of dental CAD/CAMs. Furthermore,

L. Tayebi et al., *3D Printing in Oral & Maxillofacial Surgery*,
https://doi.org/10.1007/978-3-030-77787-6_2

dental CAD/CAMs are limited to hard materials because forming is completed through milling [6, 7].

In 1972, Ciraud invented a powder deposition technique for meltable materials [8]. In a similar approach, Housholder devised a laser-assisted powder sintering method in 1979 [9]. In these methods, unlike the subtractive techniques, the fabrication is performed via an additive process—the objects are built layer-by-layer. In the late 1980s, the first additive manufacturing (AM) machines were introduced by 3D Systems Corporation. Since then, AM has found ever-increasing applications, mostly in aerospace, automotive, and medical industries [10].

AM made it possible to fabricate 3D objects of complex compositions and geometries without the need for any other manufacturing processes. Furthermore, it facilitates the construction of customized and personalized parts in the shortest possible time, with minimal material waste. Although it was originally thought that the application of AM techniques was limited to rapid prototyping, it was quickly accepted as an efficient approach for rapid manufacturing [11, 12].

Various AM-based techniques cover a much wider range of materials compared to CAD/CAM systems. Therefore, AM provides more flexibility in the fabrication of dental and maxillofacial models and prostheses.

3D printing refers to a set of AM methods that fabricate constructs in a layer-by-layer order, based on a CAD model. There might be more than 40 3D printing techniques that permit rapid manufacturing of 3D shapes with intricate details from an extensive range of metals, ceramics, polymers, composites, and biological materials [13].

This chapter will describe the five 3D printing methods with potential applications in dental and oral medicine. More specifically, stereolithography (SLA), selective laser sintering (SLS), fused deposition modeling (FDM), photopolymer jetting (PPJ), and powder binder printing (PBP) will be explicated, and their applications in oral and maxillofacial surgery will be discussed. The printable materials in four groups of metals, ceramics, synthetic polymers, and hydrogels will be introduced and their characteristics to be used in a 3D printing ink will be elaborated. The authors' strategy in this chapter is to present a preliminary introduction to the concept and methods of 3D printing by considering their applications in oral and maxillofacial surgery.

2.2   3D Printing

One can classify 3D printing techniques based on the operating principles. In some techniques, high-energy radiation (laser, ultraviolet, or electron beam) is used for polymerizing or sintering the material(s), while in others, the process is based on extrusion or jetting [14].

In the following, we will discuss the most commonly used 3D printing methods and how they can be relevant to oral and maxillofacial printing.

2.2.1   Stereolithography (SLA)

2.2.1.1 Introduction to SLA

Stereolithography (abbreviated as SLA or SL) is known as the first commercialized AM method. It was invented by Chuck Hull, the co-founder of 3D Systems, in 1986. SLA is based on the solidification of the liquid photopolymers by laser. To fabricate a part in accordance with the CAD model, laser radiation is controlled by a computer to limit the photo-induced polymerization and solidification to the desired positions [15].

The SLA process consists of the following steps (Fig. 2.1):

(a) *Laser radiation on the surface of the liquid resin*: The laser reflects the cross-sectional image of the model on the resin's surface. As a result, resin solidifies to the depth of curing and adheres to the supporting platform.
(b) *Platform movement and surface recoating*: The platform moves away from the surface of the solidified resin, and the built layer is recoated with liquid resin. To ensure the complete solidification of the resin, the platform step height is set lower than the depth of curing.
(c) Steps a and b are repeated to build all the layers.
(d) *Draining and rinsing*: The part is drained and rinsed to remove excess resin.
(e) *Post-curing*: The finished part contains unreacted groups. Postcuring with ultraviolet light completes the photopolymerization process and increases the mechanical strength of the structure [16].

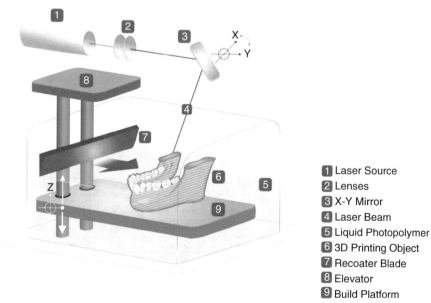

1 Laser Source
2 Lenses
3 X-Y Mirror
4 Laser Beam
5 Liquid Photopolymer
6 3D Printing Object
7 Recoater Blade
8 Elevator
9 Build Platform

Fig. 2.1 Schematic representation of the conventional SLA method

The process described in Fig. 2.1 is known as the conventional bottom-up SLA. A modified setup is also available for SLA, in which the built layer is immersed inside the liquid resin from above, and the build platform moves up vertically during the printing process. Compared to the conventional SLA, top-down SLA—the latter setup—requires less resin. In addition, if this top-down configuration is used, there is no restriction on the height of the object. However, in the former method, the height of the part is limited to the size of the resin vat. Above all, the number of pores is reduced in the final structures fabricated using the top-down setup. This is due to the removal of the surface oxygen layer, which exhibits an inhibiting effect in the conventional SLA [17].

2.2.1.2 Advantages and Disadvantages of SLA

Table 2.1 summarizes the important features of common dental SLA machines. The high vertical and lateral resolution of SLA is the main advantage of this method. The printing speed is also as high as 1.5 cm/h. This indicates that precision does not slow down the printing speed in SLA. Simultaneous high speed and precision is the reason for the popularity of this method, although it is an older technique compared to other 3D printing methods.

Despite its high speed and precision, SLA is limited to photopolymers. The parts fabricated by SLA are usually made of high-cost low-molecular-weight epoxy or acrylic resins. It is not also possible to print more than one type of resin during the printing process. Another drawback of this method is that the printed parts are brittle and subject to shrinkage after polymerization [15, 18].

The advantages of SLA have led many researchers to exploit this method for fabricating ceramic parts. To accomplish this, suspensions of liquid resins containing fine ceramic powders are required. After the printing process, the polymeric content is removed via a slow thermal treatment. At this stage, precise control of the heating rate is crucial to reduce the concentration of thermal stresses in the final structure and obtain parts with minimal structural shrinkage. Following the removal of the polymer, sintering at 500–1500 °C is performed to increase the density and enhance the mechanical properties of the ceramic body. This approach has been successfully used to fabricate dense ceramic parts, such as alumina (Al_2O_3), zirconia (ZrO_2), silica (SiO_2), and silicon carbide (SiC) [19, 20].

Table 2.1 Characteristics of common dental SLA machines (based on information published on manufacturer websites)

Model	Manufacturer	Launched in	Max. build size (xyz)	Min. layer thickness	XY accuracy	Max. printing speed
ProJet 1200	3D systems	2014	43 × 27 × 150 mm	30 μm	60 μm	14 mm/h
Ember	Autodesk	2015	64 × 40 × 134 mm	10 μm	50 μm	18 mm/h
B9CREATOR V1.2HD	B9Creations	2015	104 × 75 × 203 mm	30 μm	30 μm	NR
Form 2	Formlabs	2015	145 × 145 × 175 mm	25 μm	145 μm	10–30 mm/h
Nobel 1.0	XYZprinting	2017	128 × 128 × 200 mm	25 μm	300 μm	5–15 mm/h

2.2.1.3 Oral and Maxillofacial Applications of SLA

The oral and maxillofacial applications of SLA can be categorized as follows.

A. Fabrication of anatomical orofacial models: SLA has been employed as an invaluable technique to fabricate anatomically matched models used in reconstructive surgeries of the orofacial region. These models are most commonly used in mandibular reconstruction. For this purpose, first, a complete mandible model is made using the 3D images obtained from CT. The anatomical SLA model is then used to prepare titanium plates that direct the bone regeneration process at the site of the lesion. Titanium plates are bent in accordance with the SLA-fabricated model and are then implanted. This approach guarantees the reconstruction of large lesions with completely natural contouring. SLA-fabricated models increase the accuracy and improve the predictability of reconstructive surgeries. Furthermore, they facilitate the surgeries by reducing the operative time and eliminating the challenges related to intraoperative decision-making [21, 22].

B. Fabrication of surgical guides and templates: Placement of dental implants is challenging, especially when it comes to multiple adjacent units. In such cases, implants should be aligned exactly in parallel. Computer-aided implant surgery has been developed to achieve this ideal alignment. This approach is centered on using surgical guides and templates, which are designed and tailored based on the 3D anatomy of the patient's teeth and adjacent tissues [23].

SLA is a reliable approach to fabricate patient-specific surgical guides. The anatomic data are routinely acquired from cone-beam computed tomography (CBCT) images. SLA-fabricated surgical templates allow for the insertion of dental implants with submillimeter accuracy [24, 25].

Another application of SLA-assisted surgeries is to place orthodontic mini-implants. This approach has been used for over a decade. With the use of SLA guides, placement of the mini-implants and mini-screws has become a quick and easy process, with unprecedented safety and accuracy [26–28].

C. Fabrication of dental restorations: Fabrication of small ceramic parts with complex geometry and multifaceted hierarchical internal structure is not feasible via conventional subtractive methods. Milling tools also are unable to access the innermost sections of the ceramic bodies. SLA has successfully addressed these challenges. The progress made in the fabrication of ceramic parts by means of SLA has made it a promising method to build all-ceramic dental restorations. Regardless of the size, shape, and internal structure, SLA can fabricate dental crowns, bridges, and other prosthetic components with high accuracy and minimal material waste [29, 30].

2.2.2 Selective Laser Sintering (SLS)

2.2.2.1 Introduction to SLS

In 1986, Carl Deckard and Joe Beaman from the University of Texas at Austin invented a powder-based additive manufacturing method called selective laser sintering (SLS). SLS is performed in two ways—direct and indirect. In direct SLS, a

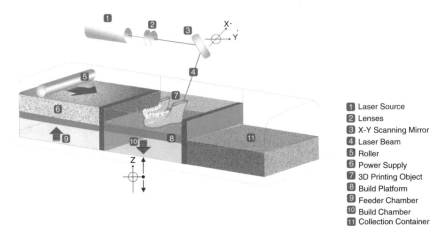

1 Laser Source
2 Lenses
3 X-Y Scanning Mirror
4 Laser Beam
5 Roller
6 Power Supply
7 3D Printing Object
8 Build Platform
9 Feeder Chamber
10 Build Chamber
11 Collection Container

Fig. 2.2 Schematic representation of the SLS method

high-power laser beam is irradiated to sinter the powder particles at the points of interest (Fig. 2.2). If the powder particles do not fuse under the laser radiation, indirect SLS would be the solution. In indirect SLS, a sacrificial polymeric binder is added to the formulation. After printing, the as-fabricated part is heated in the furnace to remove the binder from the body of the built part.

The term *selective laser sintering* is mostly used for ceramic and polymer powders. In the case of metals, some authors prefer to instead use the term *selective laser melting (SLM)*. Theoretically, SLS has no limitations on the type of materials, and a wide range of metallic, ceramic, and polymeric parts can be fabricated by this method. SLS resolution varies depending on the size of the particles and the power of the laser beam [31, 32].

2.2.2.2 Advantages and Disadvantages of SLS

Important features of some dental SLS machines are listed in Table 2.2. During the SLS, particles undergo heating and subsequent cooling processes that lead to the deformation or shrinkage of the built part. A drawback of SLS is that some molecules cannot tolerate high temperatures caused by laser radiation. For instance, biomolecules degrade at high temperatures and lose their function and, therefore, cannot be printed using SLS [33, 34].

The most significant advantage of SLS is that any material in its powder form can be printed by this method if the material withstands laser heat and shrinkage. In most polymers, ceramics, metals, and alloys, SLS is performed without the need for binding materials. Polymeric binders are used only if the powder particles are non-sinterable. Low-melting point binders make it possible to print these materials, as well. Therefore, virtually any material that can survive both laser heat and shrinkage could be printed via SLS.

Another disadvantage of this method is the need for pre- and post-heating treatments of the powdered materials used in SLS. To decrease the laser-induced

Table 2.2 Characteristics of common dental SLS machines (based on information published on the manufacturer's website)

Model	Manufacturer	Launched in	Max. build size (xyz)	Min. layer thickness	XY accuracy	Max. printing speed	Laser technology
ProX DMP 200	3D systems	2015	140 × 140 × 125 mm	10 μm	20 μm	NR	Fiber laser 300 W
Lisa	Sinterit	2015	150 × 200 × 150 mm	75 μm	50 μm	10 mm/h	IR LED 5 W
ProMaker P1000	Prodways	2015	300 × 300 × 300 mm	60 μm	NR	1.1 l/h	CO_2 30 W
VIT SLS	NATURAL ROBOTICS	2017	250 × 250 × 250 mm	50 μm	60 μm	20 mm/h	CO_2 40 W
FORMIGA P 110 Velocis	EOS	2018	200 × 250 × 330 mm	60 μm	NR	1.2 l/h	CO_2 30 W

shrinkage of the constructs, the procedure must be done at a temperature between the crystallization and glass transition or the melting points of the materials [35]. Basically, only materials that can be used in SLS are those that can endure both laser heat and shrinkage.

It is known that the quality of SLS scaffolds is very much dependent on the parameters of SLS procedures that must be optimized [36, 37]. These parameters include laser power, powdered material, scan spacing and speed, layer thickness, roller speed, and part bed temperature [36, 37]. For example, Duan et al. carefully adjusted laser power, scan spacing, and laser thickness to fabricate scaffolds consisting of poly(hydroxybutyrate-co-hydroxyvalerate) (PHBV) and calcium phosphate/PHBV with optimized mechanical/compressive properties, accuracy, and stability [36].

2.2.2.3 Oral and Maxillofacial Applications of SLS

The oral and maxillofacial applications of SLS include the following.

A. Fabrication of maxillofacial anatomic models: Maxillofacial anatomical models can be fabricated via SLS. Three-dimensional SLS models allow detailed visualization of the location and dimensions of the defects in relation to adjacent tissues. Mandibular models printed with SLS provide accurate anatomic details, with minimum deviations from dry skull dimensions and CT images. The dimensional accuracy of SLS mandibular models is higher than models printed via PPJ and PBP. Highly accurate SLS models are reliable diagnostic aids that facilitate preoperative planning and provide invaluable information needed in corrective interventions of facial defects and deformities [38, 39].

B. Fabrication of patient-specific dental implants: SLS has realized the concept of patient-specific dental implants. SLS makes it possible to fabricate threaded titanium implants with customized designs resembling the natural anatomy of teeth roots. SLS implants have high dimensional accuracy. In addition, the high density of these implants results in high strength and adequate mechanical properties. Resonance frequency analysis (RFA) and pull-out strength measurements reveal that SLS implants are suitable for immediate loading. Furthermore, results of finite element analysis indicate that the maximum micro-motions around SLS implants are significantly lower than the maximum threshold needed for successful osseointegration [40].

C. Fabrication of tissue engineering scaffolds: SLS is used for the fabrication of three-dimensional tissue engineering scaffolds. A wide range of biomaterials can be sintered via SLS to form porous scaffolds. Biodegradable and non-biodegradable polymers are often used as the matrix of these scaffolds. Using SLS, ultrahigh molecular weight polyethylene (UHMWPE), polyetheretherketone (PEEK), poly(L-lactic acid) (PLLA), poly(L-lactide–co-glycolide acid) (PLGA), and poly(ε-caprolactone) (PCL) have been successfully processed alone or in a combination of bioactive ceramic particles. These scaffolds are often used to repair bone defects. However, they could also be implemented in load-bearing applications [41–43].

2.2.3   Fused Deposition Modeling (FDM)

2.2.3.1 Introduction to FDM

In 1989, Steven Scott Crump and Lisa Crump, the cofounders of Stratasys Ltd., invented fused deposition modeling (FDM). FDM is an extrusion-based method in which the materials are melted and deposited in a layered order, according to the CAD model.

The FDM device has liquefier elements that melt both the support material and the build material. The molten materials are deposited on the base surface by the movement of the extrusion nozzle in the XY plane. Layers are arranged on top of each other by the vertical movement of the build platform (Fig. 2.3).

FDM was initially designed to build parts from materials that can be melted to form filaments. The most commonly used materials in this method are thermoplastic polymers—including nylon, polylactic acid (PLA), acrylonitrile butadiene styrene (ABS), acrylonitrile styrene acrylate (ASA), thermoplastic polyurethane (TPU), polycarbonates, and polyetherimides. It is also possible to print metals and ceramics via FDM if they are mixed with thermoplastic binders [44–46].

2.2.3.2 Advantages and Disadvantages of FDM

Table 2.3 contains the technical features of some dental FDM machines. Low cost and the ability to print multiple material types in one printing process are the main advantages of FDM. Besides, objects printed with FDM do not require any post-processing treatments or modifications.

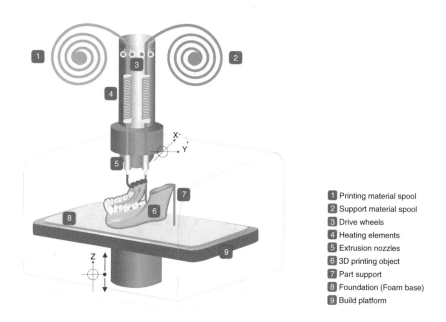

1 Printing material spool
2 Support material spool
3 Drive wheels
4 Heating elements
5 Extrusion nozzles
6 3D printing object
7 Part support
8 Foundation (Foam base)
9 Build platform

Fig. 2.3 Schematic representation of the FDM method

Table 2.3 Characteristics of common dental FDM machines (based on information published on the manufacturer's website)

Model	Manufacturer	Launched in	Max. build size (xyz)	Min. layer thickness	XY accuracy	Max. printing speed	Filament diameter
Replicator+	MakerBot	2012	295 × 195 × 165 mm	100 μm	11 μm	72 mm/s	1.75 mm
Ultimaker 2+	Ultimaker B.V.	2013	223 × 223 × 205 mm	20 μm	12.5 μm	300 mm/s	2.85 mm
LulzBot TAZ 6	Aleph objects, Inc.	2016	280 × 280 × 250 mm	50 μm	100 μm	200 mm/s	3.0 mm
Creator pro	FlashForge corporation	2016	227 × 148 × 150 mm	100 μm	11 μm	100 mm/s	1.75 mm
ORIGINAL PRUSA MK3	Prusa research	2017	250 × 210 × 210 mm	50 μm	NR	200 mm/s	1.75 mm

The most important disadvantage of FDM is the rough surface finish of the printed parts, which affects the mechanical properties and the performance of the built objects. Furthermore, cooling of the hot melt-extruded filaments usually leads to the wrapping and deviation of the finished parts from the CAD model. Several studies have been conducted to overcome these drawbacks by optimizing the printing process parameters.

In addition to the above-mentioned points, FDM is relatively slow. Moreover, the dimensions of the printable objects are limited in this method. Nonetheless, FDM is increasingly used for its versatility, cost-effectiveness acceptability, and ease of use [47, 48].

2.2.3.3 Oral and Maxillofacial Applications of FDM

The main oral and maxillofacial applications of FDM are as follows.

A. Fabrication of reconstructive mandibular models: FDM is an effective method for the fabrication of mandibular models used in planning reconstructive surgeries. Real-sized FDM models provide the possibility to study mandibular defects physically and tangibly, along with facilitating preoperational assessments and interpretations. Using FDM models decreases operational time while increasing the accuracy of surgeries. Other benefits of FDM models include improved functional and aesthetic outcomes and reduced patient stress. It is possible to build mandibular models through the use of low-cost FDM printers. Comparative studies show that these models have comparable accuracy to models printed with other commonly used rapid prototyping methods [49, 50].

B. Repair of facial fractures: Reconstruction of facial fractures is extremely important because they often cause functional and aesthetic problems. However, the existence of numerous small-sized bones with complex geometries has made facial reconstructions difficult and challenging. One of these cases is orbital blowout fracture, which is generally repaired by the use of autogenous bone graft harvested from the patient's calvaria. Prior to the implantation, the graft should be molded to fit the bone void. This could be done by the use of 3D facial models printed via FDM. FDM models allow the surgeon to refine the shape and the dimensions of the graft to fit the defect accurately. By guaranteeing the maximum dimensional conformity between the graft and the void, FDM models minimize the postoperative changes in orbital cone volume and prevent related complications, such as enophthalmos [51].

C. Fabrication of anatomic models to repair complicated cranial defects: Cranial bone defects and skull deformities are routinely repaired using acrylic implants. These implants are conventionally fabricated by moulage and mold formation methods. Fabrication of cranial implants is challenging in complex-shaped and large-sized skull defects in which accurate detection of defect margins is difficult. 3D skull models created with FDM facilitate cranioplasty surgeries by providing an accurate 3D template to repair challenging skull defects. FDM anatomic models guarantee maximum marginal conformity and anatomic contour, thus eliminating the need for several implant refinements during operation [52, 53].

D. Accurate diagnosis and treatment planning in challenging implant surgeries: Simultaneous sinus lifting and flapless implant surgery is a challenging procedure that requires accurate diagnosis and treatment planning. Conventional gypsum models only replicate the surface anatomical features and do not provide enough details to determine the appropriate number and position of implants. In addition, surgical templates fabricated on these models are not sufficiently precise, and therefore many dentists are unwilling to use them.

FDM models are reliable aids that provide invaluable details about the surrounding hard and soft tissues at the implantation site. The benefits of using these real-sized models in the diagnosis and treatment process include exact determination of the amount of available osseous tissue, determination of the amount of grafting material needed for augmentation, maximum preservation of surrounding hard and soft tissues, and minimum injuries, pain, and bleeding. Furthermore, the surgeon can rehearse the proposed procedure on the model prior to the implantation and use the precise surgical guides and drills fabricated based on the 3D model during the surgery [54, 55].

E. Fabrication of facial prostheses: Facial prostheses are conventionally fabricated via manual methods. These methods are laborious yet error-prone. In addition, the experience and skills of the technician and the anaplastologist determine the quality of the final prosthesis. A reliable alternative approach for the above-mentioned methods is FDM. As an example, FDM has been employed for the fabrication of auricular prostheses from silicon rubber. Using FDM, color-matched prostheses with high dimensional and geometrical accuracy are available in a much shorter time compared to manual methods. It is worth noting that the fabrication cost of the FDM prostheses is much lower than other 3D printing methods [56].

2.2.4 Photopolymer Jetting (PPJ)

2.2.4.1 Introduction to PPJ

Hanan Gothait, the founder and chief executive officer of XJet Ltd., introduced photopolymer jetting (PPJ) in 2001. PPJ is based on material jetting. In this technique, the photopolymers and supporting materials are jetted from an inkjet-like printhead on the build platform. Then, the materials are cured and solidified by ultraviolet radiation. The print head moves along the build platform and deposits the layers one by one (Fig. 2.4). After the completion of the printing process, the supporting materials are removed from the built part [57]. A broad range of photopolymers, rubbers, and waxes can be printed by PPJ [58].

2.2.4.2 Advantages and Disadvantages of PPJ

Table 2.4 provides the important characteristics of some dental PPJ printers. All in all, PPJ is a relatively fast method that has the ability to print objects with high resolution and quality. The cost of PPJ technology is relatively low; however, the high cost of materials restricts its application.

As already mentioned, PPJ has been adopted to print photo-polymerizable materials. Therefore, it cannot be considered as a practical way to print metallic or ceramic particles. Another drawback of PPJ is that the printed parts are mostly

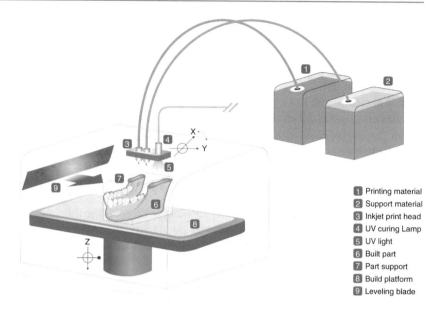

Fig. 2.4 Schematic representation of the PPJ method

Table 2.4 Characteristics of common dental PPJ machines (based on information published on the manufacturer's website)

Model	Manufacturer	Launched in	Max. build size (xyz)	Min. layer thickness	XY accuracy
3Z Lab	Solidscape	2012	152.4 × 152.4 × 50.8 mm	6 μm	25.4 μm
Objet30 Dental Prime	Stratasys	2015	300 × 200 × 100 mm	16 μm	100 μm
ProJet MJP 3600	3DSystems	2016	298 × 185 × 203 mm	16 μm	25 μm
J700 Dental	Stratasys	2017	490 × 390 × 200 mm	10 μm	100 μm
ProJet MJP 5600	3DSystems	2017	518 × 381 × 300 mm	13 μm	25 μm

thermally unstable and do not allow heat sterilization. Besides, complete removal of the support materials is not always straightforward and may cause structural damage. On the other hand, residuals of support materials may trigger skin reactions and irritation [59, 60].

2.2.4.3 Oral and Maxillofacial Applications of PPJ
The oral and maxillofacial applications of PPJ are as follows.

   A. Fabrication of teeth replica models: PPJ is an efficient method for building teeth replica models. Replicas printed via PPJ have been used in clinical diagnosis and treatment of auto-transplantation, tooth impaction, and other cases that require high accuracy. Mean deviations between PPJ replicas and natural teeth are measured at 3.8 µm, which is much less than the threshold for being considered clinically acceptable (250 µm). Comparative analyses have revealed that replicas fabricated via PPJ are more accurate and have smoother surfaces than those printed with FDM [61, 62].

B. Fabrication of full-arch dental models: PPJ is a robust method for the fabrication of full-arch dental models. Several studies have been carried out to create dental models using different 3D printing methods and compare the results. It has been demonstrated that PPJ models offer the highest accuracy compared to models printed with FDM and SLA. PPJ maxillary and mandibular models are accurate enough to be used for the rapid manufacturing of patient-specific orthodontic appliances [63].

C. Fabrication of interim crowns: PPJ has been used for the fabrication of interim crowns. Compared to crowns produced via conventional milling and compression molding methods, PPJ crowns exhibit more accuracy, especially in the occlusal region. In addition, PPJ improves the fit of the interim crowns in the internal, proximal, and marginal regions. Temporary crowns printed via PPJ protect pulpal and periodontal tissues and maintain the function and esthetics of the teeth [64].

2.2.5   Powder Binder Printing (PBP)

2.2.5.1 Introduction to PBP

Powder binder printing (PBP), also known as binder jetting (BJ), was invented by Emanuel Sachs and Michael J. Cima at the Massachusetts Institute of Technology (MIT) in 1993. In PBP, a roller levels a thin layer of the powder onto the surface. Then, the liquid adhesive (called the binder or the glue) is printed from the printing head onto the powder layer. This process continues until the full part is created (Fig. 2.5). Since the printed parts have low densities, post-processing densification is performed via light curing, sintering, or isostatic pressing. This enables to tailor the density and porosity of the final object.

PBP can be used to fabricate any powder materials. The binder binds the consecutive layers; therefore, there is no need for supporting materials [65, 66].

2.2.5.2 Advantages and Disadvantages of PBP

Important features of some PBP machines are indicated in Table 2.5. The most prominent advantage of PBP is its capability to print powder materials without the need for photopolymers or support materials. PBP uses low-cost materials and technology. Moreover, it is a relatively fast process.

The main disadvantages of PBP are related to incomplete bonding between the binder and the powder particles. First of all, it leads to porous structures and decreases the mechanical properties of the printed objects. It also increases the surface roughness of the built part. Furthermore, printed objects that have unbonded particles are low in transparency, which makes it difficult to study them using

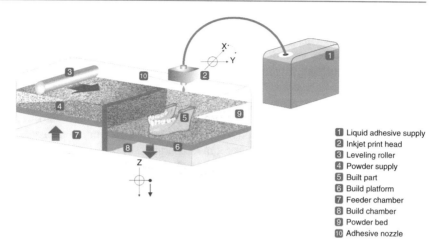

1 Liquid adhesive supply
2 Inkjet print head
3 Leveling roller
4 Powder supply
5 Built part
6 Build platform
7 Feeder chamber
8 Build chamber
9 Powder bed
10 Adhesive nozzle

Fig. 2.5 Schematic representation of the PBP method

Table 2.5 Characteristics of common PBP machines (based on information published on the manufacturer's website)

Model	Manufacturer	Launched in	Max. build size (xyz)	Min. layer thickness	XY accuracy	Max. printing speed
M-flex	ExOne (USA)	2012	400 × 250 × 250 mm	50 μm	60 μm	1–2 layers per minute
ProJet CJP 360	3D systems (USA)	2014	203 × 254 × 203 mm	100 μm	150 μm	20 mm/h
ProJet CJP 660Pro	3D systems (USA)	2015	254 × 381 × 203 mm	100 μm	100 μm	28 mm/h
ZPrinter® 450	3D systems (USA)	2015	203 × 254 × 203 mm	90 μm	NR	2–4 layers per minute
PartPro350 xBC	XYZprinting (Taiwan)	2017	350 × 222 × 200 mm	100 μm	20 μm	18 mm/h

microscopic methods. The other disadvantage of PBP is that the binders are usually bio-toxic; therefore, fabrication of tissue engineering scaffolds via PBP is limited [67–70].

2.2.5.3 Oral and Maxillofacial Applications of PBP

The oral and maxillofacial applications of PBP are classified as follows.

A. Fabrication of bone tissue engineering scaffolds: Several studies have been conducted to print synthetic bone grafts and scaffolds via PBP. Tricalcium phosphates (α and β), hydroxyapatite (HA), and combinations of them in the form of bisphasic calcium phosphates (BCPs) are the most frequently used powder materials for this purpose. In general, a dilute phosphoric acid solution is used as the binder. PBP provides an efficient method for fabricating tailored bone grafts and scaffolds with intricate internal and external geometries in compliance with the dimensions of the bone defects [71, 72].

B. Fabrication of cranial implants: PBP is an auspicious method to fabricate patient-specific skull implants. Dicalcium phosphate dihydrate (DCPD) implants are fabricated by printing liquid phosphoric acid onto tricalcium phosphate (TCP) powder. Since the process is done at room temperature, it is possible to load bioactive agents and drugs in the implants. The resultant implants offer proper cytocompatibility, controllable degradation and resorption rates, and acceptable dimensional fit [73, 74].

C. Fabrication of facial prostheses: Compared to the conventional impression techniques, PBP is a fast but accurate method to fabricate facial prostheses. Using PBP, the whole design and fabrication process is done in fewer visits with minimal stress and discomfort for the patient. This is a huge advantage, particularly in the case of pediatric patients, and minimizes the need for sedative and anesthetics agents. In addition, PBP enables control of the dimensional fit and aesthetic aspects of the final prosthesis, including color- and shade matching [75, 76].

D. Fabrication of complex-shaped prostheses: PBP is useful for the fabrication of dental and maxillofacial prostheses with multiple complex-shaped parts. Partial denture frameworks have many intricate small connectors, clasps, and retentive arms. For this reason, their fabrication using conventional subtractive methods is difficult and time-consuming. PBP has made it possible to print dental frameworks made of Ti- and Co-Cr-based alloys with unprecedented speed and accuracy. After sintering, the density of the parts reaches above 99%. Moreover, the dimensional variation of the printed frameworks from patient anatomy is negligible [77].

D. Fabrication of mandibular models: PBP has been used to reproduce mandibular anatomy. However, it has been demonstrated that the dimensional precision and anatomic details of BPB models are lower than the models fabricated via SLS or PPJ. For this reason, it is recommended to use PBP models in clinical practice with care [39, 78].

Table 2.6 provides a summary of important features of common 3D printing methods with oral and maxillofacial applications.

2.3   Printable Materials with Potential Applications in Oral and Maxillofacial Surgery

To make a 3D printed construct for oral and maxillofacial surgery, including both soft and hard constructs, appropriate printable materials with acceptable reproducibility—both in the printing procedure and the ultimate printed object—is necessary. There are four main groups of materials that are available to produce printable inks for constructs that can be used in oral and maxillofacial surgery [79]: metals, ceramics, synthetic polymers, and hydrogels. Here, we review and discuss each

Table 2.6 Comparison of common 3D printing methods with oral and maxillofacial applications

Method	Mechanism of printing	Printable materials	Min. layer thickness	XY accuracy	Max. printing speed	Need for binders	Need for supporting materials	Need for post-printing treatments	Applications
SLA	Photopolymerization of liquid resins	Liquid photopolymers	10 μm	30–300 μm	30 mm/h	No	No	Yes	Fabrication of mandibular models, surgical guides and templates, crowns, and bridges.
SLS	Laser sintering of powder particles	Powders of polymers, ceramics, metals, and alloys	10 μm	20–60 μm	20 mm/h	Yes	No	Yes	Fabrication of mandibular models, patient-specific dental implants, and tissue engineering scaffolds.
FDM	Extrusion of molten filaments	Thermoplastic polymers	20 μm	10–100 μm	300 mm/s	No	Yes	No	Fabrication of mandibular models, facial prostheses, and surgical guides and templates. Repair of facial fractures and cranial defects.
PPJ	Ultraviolet solidification of jetted photopolymers	Photo-polymerizable materials	6 μm	25–100 μm	NR	No	Yes	No	Fabrication of full-arch dental models, teeth replica models, and interim crowns.
PBP	Jetting of liquid binder onto the powder surface	Powders of polymers, ceramics and metals	50 μm	20–10 μm	28 mm/h	Yes	No	Yes	Fabrication of mandibular models, dental and maxillofacial prostheses, and crania implants.

group to understand their possible applicability in making hard and soft constructs for different conditions in oral and maxillofacial surgery. Table 2.7 summarizes the information regarding the primary printable materials in each group and their highlighted characteristics.

2.3.1   Metals

Metals have various biomedical applications, but not many of those applications take advantage of the printability of the metals [103]. Printable metals include titanium alloys, cobalt, iron, stainless steel, and chromium [104].

3D printed constructs of metals can be appropriate for use in hard tissue-related applications due to their mechanical strength [105, 106]. Among the few uses of 3D printed metal constructs in biomedical applications, one may refer to the use of 3D printed titanium alloys [80, 106], cobalt-chromium alloys [81], nitinol [82], and stainless steel [83] in various implants. An example of a 3D printed titanium scaffold is shown in Fig. 2.6. [106] The scaffolds have then been modified by deposition of HA on its surface electrochemically to induce bioactive properties to the 3D printed construct for use in bone ingrowth applications.

Besides the inertness and nondegradability of metals, their mechanical strength is the key factor that encourages their use in load-bearing applications related to oral and maxillofacial surgery.

2.3.2   Ceramics

High stiffness and bioactivity are the two essential characteristics of ceramics that have made them attractive for biomedical applications [107, 108]. Many ceramics are biocompatible and can provide a natural and osteoinductive environment appropriate for hard tissue growth [109, 110]. Thus, they are mostly used in oral and maxillofacial [111] as well as orthopedics [112] applications. HA [84], TCP [85], and HA- or TCP-based composite materials [113, 114] are the main printable ceramics, described further here:

- **Hydroxyapatite (HA):** Mimicking the configuration, mineral and mechanical properties of natural bone are the main subject of investigations on 3D printed ceramic constructs. Thus, HA, which is chemically similar to the mineral phase of bone, is one of the important ceramics used for this purpose (Fig. 2.7) [84, 115, 116].

For example, bone tissue engineering scaffolds were made by Michna et al. using an extremely concentrated HA ink by PBP or direct 3D printing [117]. The viscosity and composition of the ink were carefully adjusted in this study to 3D print self-supporting HA scaffolds with very low organic content (<1 wt %) [117].

Table 2.7 Information regarding the main printable material groups

Metals	
Examples of main printable materials	• Titanium alloys [80] • Cobalt–chromium alloys [81] • Nitinol [82] • Stainless steel [83]
Highlighted characteristics	• Mechanical strength • Inertness • Non-degradability (a few exceptions exist)
Ceramics	
Examples of main printable materials	• Hydroxyapatite [84] • Tricalcium phosphates [85]
Highlighted characteristics	• High stiffness • Brittleness • Bioactivity
Synthetic polymers	
Examples of main printable materials	• Poly(lactic acid) (PLA) [86] • Poly(caprolactone) (PCL) [87] • Poly (D,L-lactic-co-glycolic acid) (PLGA) [88] • Polyether ether ketone (PEEK) [89] • Acrylonitrile butadiene styrene (ABS) [90] • Poly(butylene terephthalate) (PBT) [91]
Highlighted characteristics	• **PLA**: Brittleness, releasing of acidic byproducts, low cost, poor cell integration • **PCL**: Low cost, slow degradation, high hydrophobicity, flexibility, biocompatible but not very good cell integration • **PLGA**: Buildup of acidic oligomers, wide range of degradability and properties based on the molecular weight ratio of LA:GA, not very good cell attachment • **PEEK**: Mechanical strength, heat resistance, inertness, risk of triggering foreign body reaction, high cost, bio-inert, radiolucent, very high melting point (~350°C) • **ABS**: Good toughness (not as brittle as polyesters), mechanical strength, non-degradability, very poor cell integration • **PBT**: Similar to PCL and PLA but with higher melting point (225°C), poor cell integration **All synthetic polymers**: Low biological activity and poor cell attachment as a result of intrinsic hydrophobicity
Hydrogels	
Examples of main printable materials	• Alginate [92] • Gelatin [93] • GelMA [94] • Hyaluronan [95] • Collagen [96] • Others (silk [97], fibrin [98], decellularized extracellular matrix [99], naturally derived matrigel [100], elastin [101], and decellularized adipose tissue [102])

(continued)

Table 2.7 (continued)

Highlighted characteristics	• **Alginate**: Nontoxic cross-linking by positively charged ions (such as Ca^{2+} or Ba^{2+} or Mg^{2+}) and the possibility of in situ cross-linking, mechanical strength, flexibility, possible deformation of the final printed construct, popular in cell-laden printing, poor cell attachment (although alginate has many desired characteristics, its biological cell integration is not as good as some other hydrogels, such as gelatin) • **Gelatin**: Generally-regarded-as-safe (GRAS) material, low-cost, good attachment to any cell type (highly biocompatible), good matrix material, suitable speed of degradation, low antigenicity, soft but relatively durable mechanical properties • **GelMA:** Photocurable (the UV cross-linked GelMA construct is not mechanically as strong as a chemically crosslinked gelatin construct), good attachment to various cell types, appropriate for cell-laden printing while inheriting many good properties of gelatin • **Hyaluronan:** Highly viscous (can be printable in a modified form or in combination with other printable hydrogels), has favorable characters in regulating the functions of cells/tissues and cell signaling • **Collagen:** Highly biocompatible, important constituent of the ECM low-immunogenic, poor mechanical properties, challenging sterilization, easily shrinkable/contractable, high-cost, extended crosslinking time, thermosensitive (causing complications in the sterilization) • **All hydrogels**: Good degree of flexibility, soft texture and supple morphology, capability of holding a good amount of water (as opposed to the hydrophobic property of polymers) while staying insoluble and keeping their 3D printed configuration

- **Tricalcium phosphates (TCP) and composites:** After HA, TCP (composed of α- and β-TCP) is the second most common calcium phosphate phase in natural bone and thus has been a focus in 3D printing scaffolds for bone repair in the oral and maxillofacial area [85, 118].

While the degradability of HA is slow, TCP can degrade faster and is often used in combination with HA to adjust the degradation [119]. The mechanical strength of TCP is low; thus, pure TCP is not appropriate for making 3D constructs. It is desired to use other elements to improve the mechanical properties of TCP by making composite materials [113]. For example, Tarafdar et al. showed that the inclusion of strontium and magnesium oxide to the 3D printed TCP scaffold can considerably improve its mechanical properties [114].

Using polymer additives with HA and TCP to make 3D printed constructs with enhanced binding property is also typical [120]. For example, in various studies, PCL as a polymer additive has been used with β-TCP to enhance the interlayer binding in the 3D printing procedure, resulting in constructs with improved mechanical properties [121, 122].

In another study, alginate has been mixed with TCP to fabricate 3D printed scaffolds with decreased brittleness and enhanced mechanical strength [123]. Phosphoric acid has been used as the binding solvent in this construct. Compared to the pure TCP 3D printed construct (the control sample), the inclusion of alginate as a natural polymer not only enhanced the mechanical property of the 3D printed construct but

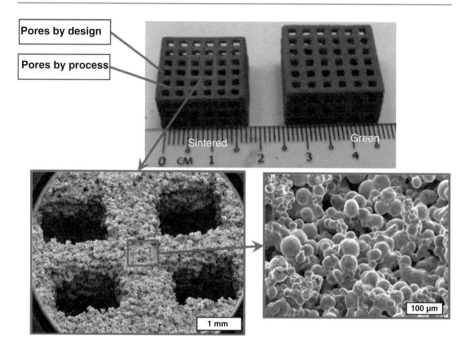

Fig. 2.6 Example of a 3D printed metal construct. The figure presents the images of 3D printed titanium with bimodal pore size distribution. Two types of pores are shown in the figure, pores by process and pores by design. The figure is obtained from reference [106] with permission

also improved the biocompatibility of the scaffolds, evidenced by increased proliferation of MG63 osteoblastic cell [123]. Collagen is another natural polymer that has been mixed with TCP to improve the properties of constructs for use as osteochondral scaffolds [124].

Due to the mechanical characteristics of ceramics, including high stiffness and brittleness, along with their bioactive properties, 3D printed constructs produced by this group of materials are appropriate for hard tissue repairs in the oral and maxillofacial region. They have no use in soft tissue regeneration related to single or multi-tissue defects in oral and maxillofacial surgeries.

2.3.3   Synthetic Polymers

Man-made polymers in three common categories of thermoplastics, elastomers, and synthetic fibers are known as synthetic polymers [125]. Various structures with different properties can be assumed for synthetic polymers to be used for a wide range of functions in biomedicine, including applications in therapeutics, drug delivery, diagnostics, and tissue engineering [125].

Synthetic polymers are an important group of materials for 3D printing in general, not specifically just for biomedical applications. This is attributed to their excellent mechanical properties, low cost, and possible degradability [46, 126].

Fig. 2.7 Example of a 3D printed ceramic construct. (**a–b**) The image of the 3D printed scaffold made with 50 wt% HA. (**c**) A pore channel. (**d**) Strut of the 3D printed scaffold. (**e–f**) Surface of the scaffold. This figure is obtained from reference [115] with permission

They can be used in the form of filaments in the FDM method, solutions in SLA technique, and beads or powders in SLS [127].

Main printable synthetic polymers include PLA [86], PCL [87], PLGA [88], polyether ether ketone (PEEK) [89], poly(butylene terephthalate) (PBT), and acrylonitrile butadiene styrene (ABS).

- **Poly (lactic acid) (PLA):** The melting temperature of PLA is around 175 °C, and thus, it is appropriate for use in melt-based 3D printing procedures. More specifically, making filaments out of this polymer for use in the FDM method is relatively easy. Such filaments can be extruded around 200–230 °C.

There are two significant problems with PLA that make it unsuitable for 3D printing of constructs that might be used in delicate oral and maxillofacial regions:

1. The release of acidic degradation by-products is an important drawback of PLA, which can cause inflammation in tissues and necrosis of cells [128]. Since the degradation happens over ester bond hydrolysis, it creates a localized decrease in physiological pH through lactic acid release.
2. Although PLA is known as a material with excellent mechanical properties among synthetic polymers, it is relatively brittle and does not have any flexibility.

Serra et al. had used the PBP or direct-printing method to print various PLA-based 3D printed scaffolds [86]. To improve its mechanical property and bioactivity, a calcium phosphate (CaP) glass, as an inorganic phase with a molar composition of 44.5P2O5–44.5Ca2- O–6Na2O–5TiO2 coded G5, and polyethylene glycol (PEG) had been combined with PLA [86]. Fig. 2.8 presents its configuration. Due to its bioactive and mechanical characteristics, this construct is appropriate for bone tissue regeneration of the oral and maxillofacial area.

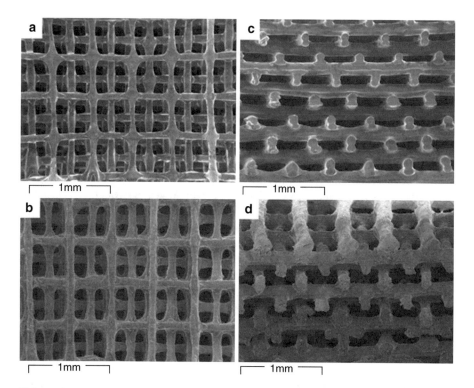

Fig. 2.8 Examples of PLA-based 3D printed constructs. Scanning electron microscopy (SEM) of 3D printed scaffolds of (**a**) PLA/PEG—top view, (**b**) PLA/PEG/G5—top view, (**c**) PLA/PEG—cross-sectional view, (**d**) PLA/PEG/G5—cross-sectional view. This figure is obtained from reference [86] with permission

- **Poly(caprolactone) (PCL):** PCL is a biodegradable, biocompatible, and low-cost polyester and thus FDM technology takes advantage of its use by making abundant PCL filaments for 3D printing [129–131]. In fact, tissue engineering and 3D printing have an effective role in making this polymer accessible and popular for biomedical applications [129].

PCL works very well in melt-based extrusion printing due to its rheological and viscoelastic characteristics and appropriate melting temperature. However, its slow degradability and stiffness make it mostly applicable for hard tissue regeneration [127]. 3D printed PCL constructs can be utilized as a support material during tissue repair, as it can stay stably and nontoxically for more than 6 months in the body and be fully resorbed in approximately 3 years [127].

PCL is one of the critical materials in the SLS method and is being employed as 10–100 μm beads [132, 133]. This technique can result in a porous scaffold with interconnected pores, good roughness, and a mechanical property comparable to the properties of bone [132]. There are many studies that indicated the successful use of 3D printed PCL for the treatment of bony defects [134, 135]. In some studies, PCL is combined with ceramics. For example, in an in vivo study, Konopnicki et al. made a 3D printed scaffold with combined PCL and β-TCP, then seeded it with porcine bone marrow progenitor cells (pBMPCs), and implanted in the 2 × 2 cm defects of minipigs' mandible. The results indicated good bone penetration depth [136].

The mechanical property of PCL allows its 3D printing in a flexible membrane shape; however, its long degradation time is not a positive factor in applications of such membranes in oral and maxillofacial surgeries. Moreover, PCL is highly hydrophobic and is not suitable for cell adhesion [137, 138]. This is an important problem with most of the synthetic polymers, which results in reduced proliferation and differentiation [139]. Also, like many other synthetic polymers, it does not have a good biological activity to facilitate tissue regeneration compared to naturally derived extracellular matrix (ECM) polymers.

Synthetic polymers, and specifically PCL, are suitable to be used for 3D printing of complex shapes, such as ears and nose [87, 140] or even as a skeleton for complex configurations needed in critical-sized defects in oral and maxillofacial surgeries. Using a sacrificial material is very common in such cases. For example, Lee et al. printed the complex configuration such as the human ear using PCL (Fig. 2.9) [87]. In this work, PEG was employed as the sacrificial material and was stacked at the same level of the PCL 3D printed layers. After completing the fabrication, PEG was simply removed from the construct by dissolving in distilled water (in an incubator for 40 min) [87]. A similar method can be used to make the complex structure that might be needed in the regeneration of specific defects in the oral and maxillofacial region.

- **Poly (D,L-lactic-co-glycolic acid) (PLGA):** PLGA is a biocompatible copolymer of PLA and poly(glycolides) (PGA) [141]. The primary use of PLGA in biomedical applications is its suitable function as a carrier for controlled drug delivery [141]. Thus, we may consider its practical application in the treatment

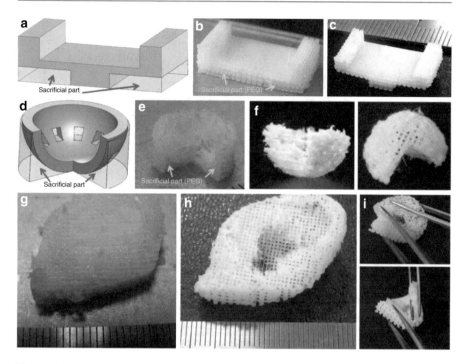

Fig. 2.9 3D printed complex structures are possible to be made by PCL. PEG was used as the sacrificial material. Panels (**a–c**) present an overhanging assembly with the size of 2 × 1 × 0.6 cm; (**a**) CAD design, (**b**) 3D printed construct with PCL and PEG, (**c**) Final structure of the overhanging shape. Panels (**d–f**) demonstrate a hollowed hemisphere shape with a diameter of 1 cm; (**d**) CAD design, (**e**) 3D printed constructs before removing PEG, (**f**) Final structure after removing PEG. Panels (**g–i**) illustrate a framework of an ear, (**g**) 3D printed structure with PCL and PEG, (**h**) The ear-shaped framework after removing the PEG, (**i**) A demonstration of the flexibility of the construct. Images are obtained from various figures in reference [87] with permission

of oral and maxillofacial defects when a specific drug is needed to be delivered and sustainably released in the area of defects. The degradation of PLGA can be varied from 24 hours to several years, depending on the molecular mass distribution and crystallinity of these elements, along with the molecular weight ratio of lactic acid and glycolic acid (LA:GA) [142].

Different PLGA compositions have been 3D printed by various methods [143–145]. For example, Mironov et al. 3D printed PLGA constructs by extrusion of the PLGA solution in tetraglycol of specific viscosity (Fig. 2.10) [146]. Cultures of line NIH 3T3 mouse fibroblasts, mesenchymal stem cells of human adipose tissue, and floating islet cultures were used to observe the in vitro cytotoxicity and cytocompatibility of this scaffold, and positive results were achieved [146].

In an extensive recent study by Guo et al., 3D printed PLGA constructs with a different molecular weight ratio of LA:GA (with no additional ingredients) were fabricated by employing a PBP or direct 3D printing technique [88]. The

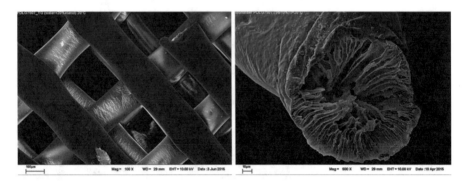

Fig. 2.10 3D printed PLGA construct: Structure of the 3D printed scaffold (left) and an individual filament (right) using SEM imaging. The figure is obtained from reference [146] with permission

dependency of the printing parameters to the PLGA compositions, the effect of the compositions on water adsorption, mechanical properties, and degradation of the 3D printed scaffolds were thoroughly examined in this study [88].

The major disadvantage of PLGA, which is highly unfavorable for its possible use in delicate biomedical applications, is having the potential trigger for the build-up of acidic oligomers, leading to inflammatory responses in the tissues [147]. Although inflammation is important in the regeneration of tissues, it is vital to control the inflammatory responses [148]. Excessive inflammation can cause fibrosis, which may distract the function of tissues or may lead to rejection of the implanted materials. This is because the inflammatory response, which is related to the immune system, can stimulate cells—such as neutrophils and monocytes—to move into the injured region of the body [149].

- **Polyether Ether Ketone (PEEK):** PEEK is known as a semicrystalline polymer [150], that can be utilized to 3D print customized craniofacial implants [151]. PEEK is usually used for bone replacement, as it is bioinert, radiolucent, biocompatible, and robust with a mechanical property similar to cortical bone [127]. It has a very high melting point (~350 °C) [152] and thus cannot be printed with usual extrusion-based printing. Only the SLS method is popular for 3D printing this material [151, 153, 154]; however, there are limited reports about using FDM method to 3D print PEEK [155, 156]. PEEK has excellent heat resistance, and therefore, steam sterilization can be applied to it [157]. Due to the lack of ability to integrate with body tissues, the risk of triggering foreign body reaction—such as dislodging, encapsulation, and extrusion—is very high for PEEK [127]. 3D printed PEEK implants are also more expensive than other polymer constructs [127].

The abovementioned characteristics of PEEK make it appropriate to be used in the fabrication of 3D printed scaffold and implants for oral and maxillofacial surgery.

- **Acrylonitrile butadiene styrene (ABS):** ABS is a triblock and petrochemical-based copolymer with a melting point of 105 °C. As it is composed of acrylonitrile and butadiene elements, its mechanical strength is outstanding [158, 159]. While polyester synthetic polymers suffer from brittleness, ABS possesses good toughness due to its styrene units [158]. It has been shown that blending poly (l-lactide) (PLLA) with ABS can increase the toughness of PLLA [160].

  FDM and SLA 3D printing systems take advantage of this synthetic polymer [158, 159, 161, 162]. However, since ABS is nondegradable and very poor in cell integration, it is not often being used as scaffold or membrane materials for biomedical applications such as for oral and maxillofacial surgery. There are limited reports of using this material in tissue engineering of bone, cartilage, and other tissues [90, 163, 164].

- **Poly(butylene terephthalate) (PBT):** PBT, as a thermoplastic polyester, is very similar to PLA and PCL [127]. Although it is as biocompatible as PLA and PCL, its melting point is very high (225 °C), which is a significant drawback for its use in 3D printing [127]. More specifically, since it does not seem to have an advantage over these two polyesters (PCL and PLA), its higher melting point makes it not very popular in biomedical applications.

  PBT can be used in FDM 3D printing systems. Tellis et al. reported the use of 3D printed PBT scaffold, made by FDM method, for bone tissue engineering of canine trabecular bones [91]. The fabricated 3D printed scaffold matched well with the trabecular bone in terms of porosity; however, its biocompatibility with cells was not examined in this study [91].

  PBT has also been utilized as coatings for scaffolds. For example, PBT-coated CaP scaffolds were appropriate for bone regeneration in a canine knee model [165]. Coating of a combination of PBT and poly(ethylene oxide) (PEO) on titanium alloy implants has also demonstrated effectiveness in bone binding [166], although it is not clear whether the positive effects are achieved because of the coating itself or the combination of coating and scaffold materials.

2.3.4   Hydrogels

Membranes have a specific position in oral and maxillofacial surgery for different applications, such as soft tissue repair and regeneration or in Guided Tissue Regeneration/Guided Bone Regeneration (GTR/GBR) procedures.

Nontoxic hydrogels are the most popular material being used to make 3D printed membranes for actual biomedical applications due to two significant characteristics [167]:

1. Hydrogels are made of 3D polymer networks with a reasonable degree of flexibility, soft texture, and supple morphology, which are beneficial characteristics to produce a flexible membrane.

2. Hydrogels are known as gels with water as their swelling agent. Thus, they are capable of maintaining the right amount of water, while staying insoluble and maintaining their 3D printed configuration, which allows them to closely mimic body tissues in terms of being hydrated constructs (in opposite to the hydrophobic property of polymers).

Therefore, hydrogels are the main focus in creating desired 3D printed membranes that can be used in oral and maxillofacial applications.

Extrusion-based printing and, to a lesser extent, SLA are known as the major methods of 3D printing hydrogels [127]. An important step in maintaining the integrity of 3D printed hydrogel constructs is cross-linking. Constructs made by chains of natural or synthetic polymer networks in hydrogels can be cross-linked chemically (e.g., covalent bonds), physically (e.g., hydrogen bonds, hydrophobic/hydrophilic associations, or ionic complexation), or by mixed chemical and physical cross-linking mechanisms [167, 168].

The major printable hydrogels are explained below:

- **Alginate:** As a plentiful natural polymer in nature, alginate can be extracted from marine brown algae (Phaeophyceae) [169]. As a structure-forming element in the brown algae and its gel matrix, it provides the plant with both flexibility and mechanical strength [170], characteristics that are desired in making suitable biomedical membranes.

Alginate, as an anionic polysaccharide, can absorb water to the extent of 200–300 times its weight [171]. Moreover, alginate is chemically similar to the ECM, and thus nutrients and metabolites can quickly diffuse into it. While sodium alginate is water-soluble in water, calcium alginate does not dissolve in water; thus, it is possible to have sodium alginate extruded from a 3D printing nozzle to a solution of calcium ion solution [171]. Basically, the most exciting characteristic of alginate as a polycation is the capability of being crosslinked using positively charged ions, such as Ca^{2+} in $CaCl_2$ [171] or Ba^{2+} [172] or Mg^{2+} [173]. This is why the use of alginate is prevalent in cell-laden printing or any 3D printing method/construct in which cell encapsulated hydrogel is somehow involved.

There are numerous reports about how various cells are encapsulated in hydrogels and used in 3D printed constructs. In some reports, the scaffolds are made with different materials, and then, cell encapsulated alginate was incorporated into the 3D printed scaffolds [174–176]. For example, chondrocyte-encapsulated alginate was incorporated in a specific SLS 3D printed scaffold made with acrylated trimethylane carbonate/trimethylolpropane (TMC/TMP) and implanted in mice [174].

In many other reports, alginate has been used in cell-laden printing to regenerate different tissues [177]. For example, using the inkjet 3D printing method, Huang et al. made zig-zag tubes with a combination of alginate and cells for use in tissue engineering of blood vessels [178]. Alginate has been used for bioprinting of blood vessels by other investigators, as well [179–181].

Alginate has also been employed for skin bioprinting [182] or in cell-laden printing with chondrocytes for cartilage tissue engineering applications [183]. Although in all of the abovementioned applications, the constructs may look like a membrane, they are mostly made to hold the designated cell for a limited time of the experiment (rather than storage for future use). Such cell-laden printed membranes are not mechanically strong enough for surgical handling [182, 183].

On the other hand, while alginate is suitable for cell-laden printing among the hydrogels, due to its nontoxic cross-linking procedure, the cell attachments of many other hydrogels—such as gelatin—are much better than alginate. This is why cell-free alginate membrane is not a perfect choice for offering prolonged cell attachment and growth [184].

Moreover, in our experimental works, we realized that alginate, even at high concentrations, tends to flow once printed on the platform. This characteristic causes deformation of the final printed object. Such deformation may not be necessary for cell-laden printing but is not favorable for having a robust cell-free membrane to be used in some applications of oral and maxillofacial surgery in which good surgical handling is a required property.

Most of the studies on alginate suggest in situ cross-linking while the object is being printed [92]. Our experience indicated that care must be taken not to cross-link the last printed layer before printing the next one on top of it, since it abolishes the adhesion between layers and, consequently, damages the object coherence. To address such a problem, in situ cross-linking is carried out by precise control of cross-linker diffusion throughout a grid used as the platform. Such a strategy requires very complicated designs. Additives can be used to adjust the viscosity and donate a liquid–solid transition to such inks for addressing this issue.

In a recent study performed by Heo et al., the membrane of alginate conjugated to bone formation peptide-1 (BFP1) was 3D printed (Fig. 2.11) [92]. The in vitro and in vivo studies were conducted on the 3D printed constructs to demonstrate that the scaffold could safely accommodate human adipose-derived stem cells (hAD-SCs) and is appropriate for bone regeneration [92].

- **Gelatin:** Gelatin is a denatured form of collagen and is achieved by hydrolyzing collagen from the animal using acid or alkaline. Gelatin is one of the safest and most reasonably priced biomaterials to be used in pharmaceuticals, medical/food products, and cosmetics [185, 186]. United States Food and Drug Administration (FDA) considers gelatin as a generally-regarded-as-safe (GRAS) material [187].

Among the many advantages of gelatin, the below ones are important for applications on oral and maxillofacial surgery:

1. It is biocompatible and low-cost with a reasonable speed of degradability [185, 186].
2. Chains of gelatin are rich in motifs, such as arginine-glycine-aspartic (RGD) sequences, which can modulate cell attachment. It can effectively attach to all

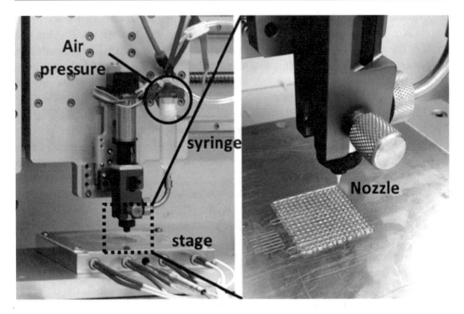

Fig. 2.11 3D printed membrane made by alginate–BFP1 (right) and the 3D printing machine used for its fabrication (left). This figure is obtained from reference [92] with permission

cell types. This is an essential advantage of gelatin compared to many other materials without cell-recognition sites, such as polymers [188].

3. Gelatin can act as a suitable matrix material, which is critical for applications in tissue engineering. Basically, cell adhesion can be supported by gelatin without involving cell phenotypes [188–190] and as a porogenic element gelatin offers structural support when combined with other material components [191].

4. Since gelatin is a denatured product, it is less antigenic than collagen [187].

5. Gelatin has soft but durable mechanical properties [192].

All these advantages make gelatin a favorable candidate for use in the fabrication of constructs that can be used in oral and maxillofacial surgery.

For 3D printing of gelatin, the dependency of its so-gel transition to temperature is the crucial player in the entire procedure of printing. Premature gelation of gelatin during printing can be the reason for the challenging 3D printing route, as this material can cause the fabrication of low-quality constructs [193]. Combining gelatin with other polymers can make the printing procedure much easier [181, 194]. While it can be costly and not be applicable for all 3D printing machines to 3D print the pure or modified gelatin, it is recommended to conduct specific adaptations on the 3D printer to improve the handling of gelatin hydrogels [94].

Laronda et al. have recently reported the 3D printing of microporous hydrogel scaffolds to examine how the pore geometry—changed by altering the angle between layers—influences the ovarian follicles (Fig. 2.12) [195]. They chemically

Fig. 2.12 A 3D printed gelatin membrane with five layers and homogamous distribution of pore. Insert presents the same membrane with higher magnification. The membrane is cross-linked by EDC/NHS. Scale bar is 250 mm. This figure is obtained from reference [195] which is licensed under a Creative Commons Attribution 4.0 International License, and no permission is required

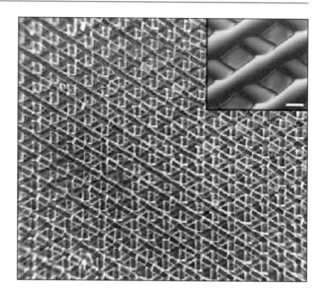

cross-linked the scaffolds using 1-ethyl-3-(3-dimethylaminopropyl)carbodiimide hydrochloride (EDC)/N-hydroxysuccinimide esters (NHS).

Shortly after this report, Lewis et al. reported 3D printing of gelatin in the various repetitive geometries (Fig. 2.13) [93]. They studied the effect of different pore geometry on the biological response of seeded undifferentiated hepatocyte cell line (HUH7) as a model hepatocyte system [93].

Cell-laden printing with gelatin has also been reported by using microbial transglutaminase (mTgase) as a cross-linker [196]. However, most investigators prefer to use photocurable methacrylated gelatin for cell-laden printing instead of gelatin itself, so they can employ UV for cross-linking, which is safer than other methods for cells [197].

In summary, the thermoreversible gelation of gelatin must be exploited to obtain a solid object after printing. The printed object must then be cross-linked to permanently fix the final shape. Various additives to the gelatin matrix can control different characteristics of the ink and the printed object. These additive options include:

1. **Thickeners**: A particular concentration of this group of materials can assist not only in adjusting the viscosity of the ink but also in modulating the gelation temperature of gelatin. Carboxymethylcellulose is an example of such additives [198].
2. **Filler to improve mechanical properties**: Various micro- and nanoparticles can be added to gelatin to enhance the mechanical properties of the final products. HA and TCP are examples of such fillers. Many organic/inorganic fillers have been used to improve the biological, mechanical, and/or degradation behavior of hydrogel scaffolds in vitro and in vivo by preparing hydrogel-based composites [199], among which calcium phosphates are an important group. This group of

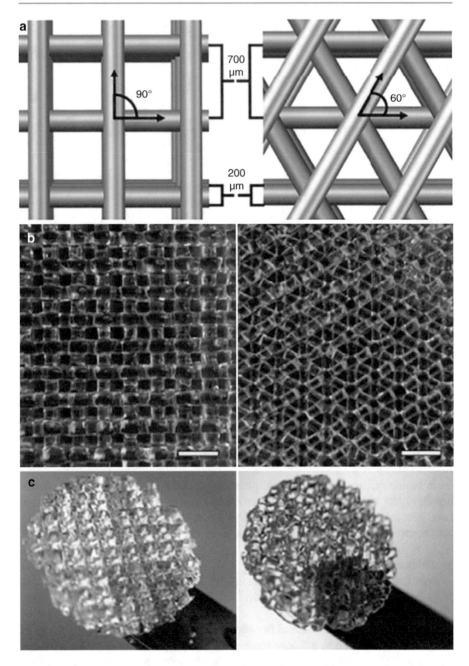

Fig. 2.13 Gelatin 3D printed membrane/construct in various repetitive geometries. (**a**) Interior pattern of the 3D printed models. (**b**) The photo of the 3D printed gelatin structure with the nozzle diameter of 0.2 mm, scale bars = 1 mm. (**c**) Cross-linked structure of the 3D printed gelatin (diameter is 6 mm). The figure is obtained from reference [93] with permission

fillers may improve bone bonding [200, 201] and in some studies reported to enhance mechanical properties [202].

3. **Other materials, proteins, and biomolecules to improve biological response**: Gelatin itself generally represents ideal cell attachment and biological response. However, different proteins could be added to the gelatin matrix to improve biological response. It becomes more highlighted when the cells of the target tissue are considered responsive to a specific protein. For example, elastin-derived motifs have been used as nonintegrin adhesion ligands in tissue-engineered hydrogels [203].

It should be noted that one or more additives could be included in the formulation simultaneously, and a single additive might have one or more functions. For example, the inclusion of hyaluronic acid to gelatin in our formulation adjusted the viscosity and improved mechanical and biological characteristics of the gelatin ink [204].

As another example, in a study performed by Shi et al., silk fibroin and bone marrow stem cells (BMSC)-specific-affinity peptide were added to gelatin to improve the 3D printed scaffold both structurally and functionally to be used for cartilage tissue regeneration [205].

To make an appropriate 3D printed construct, chitosan was also incorporated into gelatin to take advantage of the antimicrobial property of chitosan and its capability to trigger hemostasis [206]. About the interaction between chitosan and gelatin, it should be considered that gelatin is negatively charged at the pH over the isoelectric point (pHiso = 4.7) [207]. This leads to the creation of a polyelectrolyte complex due to the reaction of ammonium ions of chitosan (positively charged) with the carboxylate groups from the ampholytic gelatin. Hence, in this study, gelatin was improved with chitosan to make polyelectrolyte gelatin-chitosan (PGC) hydrogels, which were used in 3D printing of membranes suitable for skin regeneration and wound healing (Fig. 2.14) [206]. The 3D printed membranes exhibited good configurational integrity and acceptable biocompatibility using fibroblast skin cells [206].

- **Gelatin Methacrylate (GelMA):** The peptide chains of gelatin contain lysine, and this lysine terminal amino group has the ability to interact with methacrylamide and make a methacrylate group [208]. This modified version of gelatin is known as gelatin methacrylate (GelMA), which is a photocurable hydrogel that can be cross-linked with UV, rather than the usual chemical cross-linkers, such as EDC/NHS [208]. Thus, GelMA is appropriate for cell-laden printing, as cells cannot tolerate the chemical cross-linkers. However, it should be noted that the UV cross-linking is not as strong as chemical cross-linking, and thus, the 3D printed constructs made with GelMA are not mechanically as strong as the gelatin constructs and do not have secure surgical handling property. They do, however, still inherit the valuable properties of gelatin, such as biocompatibility, perfect attachment to various cell types, and biodegradability (can be absorbed in vivo in about 2 months) [171, 209]. As an example of GelMA 3D printing,

Fig. 2.14 A 3-layered 3D printed membrane made of 5% PGC hydrogels. Scale bars are 5 mm. This figure is obtained from reference [206], which is licensed under a Creative Commons Attribution 4.0 International License, and no permission is needed

Billiet et al. pre-seeded the HepG2 (liver hepatocellular carcinoma) cell line into GelMA and 3D printed the hydrogel/cell mixture by extruding it through a nozzle [94]. Shortly after printing, the 3D printed construct was cured using UV light. In such printing, the viability of cells under the extrusion pressure is an important concern, which was discussed in this study [94]. They showed that conical needles were more appropriate than cylindrical ones for such cell-laden printing [94].

• **Hyaluronan:** Hyaluronan (or hyaluronic acid), as an ECM material, can be found in any connective tissue of our body. This natural glycosaminoglycan is highly compatible for biomedical and tissue engineering applications and has already been used in the clinic for a long period of time [210, 211]. The importance of hyaluronan is related to its involvement in regulating cells and tissue functions, such as cell differentiation, migration, angiogenesis, and proliferation [210].

However, unmodified hyaluronan can be highly viscous and not easily printable (or not stable after possible printing) [95]. But, due to its favorable characters in regulating cells, it has been used in compositions with other printable hydrogels, such as GelMA [212], photocurable dextran [213], and via conjugation with thermoresponsive poly(N-isopropylacrylamide) [214]. The other approach of using this hydrogel in 3D printing is chemically modifying it [210, 215].

For example, methacrylated hyaluronan, which is photocurable, was used as one element of a multi-component ink for a 3D printing process [216–218].

Skardal et al., in a series of papers, showed the ability to use different modified hyaluronan in 3D printing inks, such as using methacrylated hyaluronan with gelatin in two-step bioprinting [218], using a composition of poly(ethylene glycol) and thiolated hyaluronan for 3D printing of vessel-like constructs [219], and using gold nanoparticle-cross-linked hyaluronan-gelatin composite in an ink [220].

In another study by Ouyang et al., the authors introduced a new hyaluronan-based ink that meets the criteria for printability in an extrusion-based 3D printing method [95]. More specifically, they coupled adamantane (Ad, guest) and β-cyclodextrin (CD, host) moieties to make a guest-host bonding, resulting in a supramolecular assembly of two hydrogel precursors [95]. Photo-cross-linkable methacrylate groups have also been used in the macromers. The authors investigated how different types of cross-linking—including guest-host, photo-cross-linking, or a grouping of these two—can affect the printability of hyaluronan to fabricate multilayer membranes [95].

- **Collagen:** Collagen is the important constituent of the ECM and one of the most plentiful proteins in the body [221]. About 25% of our dry weight is composed of collagen [221, 222]. Although more than 20 different types of collagen exist, fibril forming collagen type I is the more frequently used collagen in biomedical and tissue-engineering applications, which can be extracted from rat tail, tendons, bovine, and pork skin [221].

Collagen is biocompatible, low-immunogenic, and can be degraded by collagenases [221–223].

However, if no extra cross-linking applies to collagen, its mechanical properties will be very poor. For example, 1–5 mg/mL collagen gel type I (extracted from a rat tail) at 37 °C has a storage modulus below 100 Pa [224]. To improve its mechanical properties, collagen can be cross-linked, either chemically or enzymatically, or blended with other polymers to make double networks [225]. Some cross-linking strategies induce cytotoxic challenges for collagen to be used in biomedical applications [225]. There are also complications regarding the sterilization of collagen because of its thermosensitivity and degradation, as the temperature can change the fibrillar form of collagen [221]. Moreover, shrinkage/contraction is frequently reported in collagen scaffolds after cell activity [226–228].

Despite the shear-thinning properties of collagen, its gelation temperature, and pH dependency make it an interesting hydrogel for 3D printing. Due to its poor mechanical property, it is usually printed in preformed molds, such as 3D freeform fabrication, or as support inside various biomaterials [96, 229, 230]. It must also go through additional crosslinking after the 3D printing procedures [230–234]. 3D printing of collagen with other materials is very common, and many investigators reported such compositions for 3D printing [98, 229, 235–245].

The high cost of collagen is a barrier in using this material in 3D printer with large-capability cartridge (e.g., 30 mL). We found it very difficult and expensive to perform collagen printing, especially for fabricating large constructs. However, its use in cell-laden printing and/or in combination with other materials is reasonable [234, 242]. Various applications are considered for the 3D constructs/membranes composed of collagen. Examples include mimicking oral mucosa, skins, and grafts [231, 234], and making scaffolds for regeneration of bone [74, 229], that can be potentially used in oral and maxillofacial surgery.

2.3.5   Other Materials

Among other relevant 3D printing hydrogels, one may refer to silk [97], fibrin [98], decellularized ECM [99], naturally derived Matrigel [100], and decellularized adipose tissue [102], all of which are mostly used for cell-laden printing. Elastin is also a main ECM protein in arteries [246]. However, we hardly found any report of 3D printing of this material for biomedical applications, other than the gelatin/elastin/sodium hyaluronate membranes [101]. Elastin is usually used for the fabrication of engineered tissues in combination with collagen [247].

Silk and fibrin, as important structural proteins, are used more often than the above-mentioned materials [248]. Regarding the 3D printing of silk, long-term solution precipitation is one of the major issues in printing native silk fibroin [249, 250]. Moreover, the metastable state of silk is highly problematic in its printing [249, 250], and the addition of some additives can be helpful to resolve this issue [249, 251–253]. Making recombinant silk for 3D printing is also effective in terms of printability due to its shear-thinning behavior [250]. 3D printed inks composed of silk material have been examined in different applications, including cartilage regeneration [97], optical waveguides [254], testing cell-matrix interactions [255], small-scale catalytic motors [256], biosensors [249], strain gauges for biological use [257], and general tissue engineering [97, 255, 258].

Fibrin is an expensive FDA-approved material. The gels of fibrin have low viscosity, quick gelation, and well-behaved cross-linking procedure; however, the accuracy of 3D printing with fibrin is reported to be low (~52%) [98]. Since it has a quick gelation time and a high rate of degradation, the inks made based on fibrin are often blended with other materials, such as gelatin [259] or gelatin/alginate [260–262].

Matrigel is a commercialized blend of proteins extracted from mouse sarcoma cells and composed of laminin, collagen, entactin, heparan sulfate proteoglycans, and some growth factors [263]. The compositions of decellularized ECM depend on the living tissues they are extracted from as well as the use of the decellularization protocol [264]. However, they usually contain structural proteins, polysaccharides, glycosaminoglycans, and biofunctional factors [264].

Although Matrigel and decellularized ECM are similar to the natural matrix [99], their use in 3D printing is limited. This is mostly due to the challenges in decellularization processes and standardization of extraction materials caused by batch variability, which highly influence the reproducibility and controllability of printing [248].

Since using 1-ethyl-3-(3-dimethylaminopropyl)carbodiimide hydrochloride (EDC) and N-Hydroxysuccinimide (NHS) is a common and popular method of cross-linking in 3D printed hydrogels constructs, here we explain the mechanism of EDC/NHS cross-linking.

As shown in Fig. 2.15, during the cross-linking procedure, when EDC is included in the hydrogel solutions, it first reacts with the groups of carboxylic acid and makes

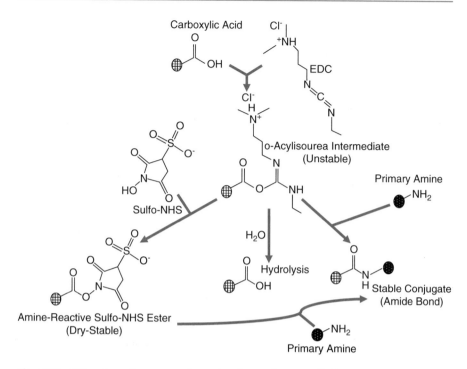

Fig. 2.15: This schematic presents the carboxyl-to-amine cross-linking reaction of EDC and Sulfo-NHS (water-soluble analog of NHS) that is frequently used in cross-linking of 3D printed hydrogels. Inclusion of NHS leads to formation of amine-reactive intermediate, which is dry-stable and, therefore, can enhance the final efficacy

an amine-reactive O-acylisourea intermediate, which is an unstable element [265, 266]. This compound impulsively reacts with primary amines and makes an amide bond and a by-product as a soluble urea derivative. Due to the instability of O-acylisourea intermediate, if it does not react with an amine, hydrolysis of the intermediate and regeneration of carboxyl can occur, and N-substituted urea will be released. Thus, quenching the EDC activation reaction by a thiol-containing compound, such as 2-mercaptoethanol, is essential [265].

Here, the role of NHS or Sulfo-NHS (water-soluble analog of NHS) is important. EDC links NHS to carboxyls, making an NHS ester that is stable and allows the effective conjugation to primary amines when the pH is appropriate (physiologic pH) [265].

EDC is known as a zero-length carboxyl-to-amine crosslinker because it can make a direct conjugation of carboxylates to primary amines [266, 267]. More specifically, after the cross-linking procedure, ECD will not be a portion of the final amide bond among the conjugated molecules [266]. EDC and its by-product, isourea, are water-soluble and dissolve in the aqueous solution. This helps the purification of the final cross-linked products [265].

2.4   Summary

In recent years, 3D printing has found increasing use in dentistry and maxillofacial surgeries. Diverse modalities of this emergent technology make it possible to fabricate the most complicated appliances and instruments, with the possibility of choosing the type of materials. In addition, 3D printing has been a reliable way to manufacture anatomical models with unprecedented dimensional precision, which significantly reduces the surgeon error rate as well as patient stress, pain, and discomfort. 3D printed surgical aids and guides also make the surgeries safe, secure, and minimally invasive. Above all, 3D printing has realized the dream of patient-specific implants and prostheses. Innovative solutions based on this technology are expected to improve the life quality of millions around the world in the near future. This chapter explained different 3D printing techniques that can be used for oral and maxillofacial applications followed by introducing various printable materials in four groups of metals, ceramics, synthetic polymers, and hydrogels that can be employed for oral and maxillofacial surgery.

References

 1. Kalpakjian S, Schmid SR, Musa H. Manufacturing engineering and technology: hot Processe. China Machine Press; 2011.
 2. Amirouche FM. Principles of computer-aided design and manufacturing. Pearson Prentice Hall; 2004.
 3. Sittig DF, Ash JS, Ledley RS. The story behind the development of the first whole-body computerized tomography scanner as told by Robert S. Ledley. J Am Med Inf Assoc. 2006;13(5):465–9.
 4. Hutchinson J, Edelstein W, Johnson G. A whole-body NMR imaging machine. J Phys E Sci Instrum. 1980;13(9):947.
 5. Duret F, Blouin J-L, Duret B. CAD-CAM in dentistry. J Am Dent Assoc. 1988;117(6):715–20.
 6. Miyazaki T, Hotta Y, Kunii J, Kuriyama S, Tamaki Y. A review of dental CAD/CAM: current status and future perspectives from 20 years of experience. Dent Mater J. 2009;28(1):44–56.
 7. Poticny DJ, Klim J. CAD/CAM in-office technology: innovations after 25 years for predictable, esthetic outcomes. J Am Dent Assoc. 2010;141:5S–9S.
 8. Ciraud PA. Process and device for the manufacture of any objects desired from any meltable material. FRG Disclosure Publication 2263777; 1972.
 9. Housholder RF. Molding process. Google Patents. 1981.
10. Zhai Y, Lados DA, LaGoy JL. Additive manufacturing: making imagination the major limitation. JOM. 2014;66(5):808–16.
11. Wong KV, Hernandez A. A review of additive manufacturing. ISRN Mech Eng. 2012;1:1–10.
12. Huang Y, Leu MC, Mazumder J, Donmez A. Additive manufacturing: current state, future potential, gaps and needs, and recommendations. J Manuf Sci Eng. 2015;137(1):014001.
13. Chua CK, Leong KF. 3D printing and additive manufacturing: principles and applications (with companion media pack) of rapid prototyping. 4th ed. World Scientific Publishing Company; 2014.
14. Bikas H, Stavropoulos P, Chryssolouris G. Additive manufacturing methods and modelling approaches: a critical review. Int J Adv Manuf Technol. 2016;83(1–4):389–405.
15. Melchels FP, Feijen J, Grijpma DW. A review on stereolithography and its applications in biomedical engineering. Biomaterials. 2010;31(24):6121–30.
16. Bártolo PJ. Stereolithography: materials, processes and applications. Springer; 2011.

17. Gross BC, Erkal JL, Lockwood SY, Chen C, Spence DM. Evaluation of 3D printing and its potential impact on biotechnology and the chemical sciences. ACS Publications; 2014.
18. Harris RA, Hague RJ, Dickens PM. The structure of parts produced by stereolithography injection mould tools and the effect on part shrinkage. Int J Mach Tools Manuf. 2004;44(1):59–64.
19. Halloran JW. Ceramic stereolithography: additive manufacturing for ceramics by photopolymerization. Annu Rev Mater Res. 2016;46:19–40.
20. Dehurtevent M, Robberecht L, Hornez J-C, Thuault A, Deveaux E, Béhin P. Stereolithography: a new method for processing dental ceramics by additive computer-aided manufacturing. Dent Mater. 2017;33(5):477–85.
21. Mehra P, Miner J, D'Innocenzo R, Nadershah M. Use of 3-d stereolithographic models in oral and maxillofacial surgery. J Maxillofac Oral Surg. 2011;10(1):6–13.
22. Gil RS, Roig AM, Obispo CA, Morla A, Pagès CM, Perez JL. Surgical planning and microvascular reconstruction of the mandible with a fibular flap using computer-aided design, rapid prototype modelling, and precontoured titanium reconstruction plates: a prospective study. Br J Oral Maxillofac Surg. 2015;53(1):49–53.
23. D'haese J, Ackhurst J, Wismeijer D, De Bruyn H, Tahmaseb A. Current state of the art of computer-guided implant surgery. Periodontol. 2017;73(1):121–33.
24. Ersoy AE, Turkyilmaz I, Ozan O, McGlumphy EA. Reliability of implant placement with stereolithographic surgical guides generated from computed tomography: clinical data from 94 implants. J Periodontol. 2008;79(8):1339–45.
25. Bencharit S, Staffen A, Yeung M, Whitley D III, Laskin DM, Deeb GR. In vivo tooth-supported implant surgical guides fabricated with desktop stereolithographic printers: fully guided surgery is more accurate than partially guided surgery. J Oral Maxillofac Surg. 2018;76(7):1431–9.
26. Kim S-H, Kang J-M, Choi B, Nelson G. Clinical application of a stereolithographic surgical guide for simple positioning of orthodontic mini-implants. World J Orthod. 2008;9(4):371–82.
27. Morea C, Hayek JE, Oleskovicz C, Dominguez GC, Chilvarquer I. Precise insertion of orthodontic miniscrews with a stereolithographic surgical guide based on cone beam computed tomography data: a pilot study. Int J Oral Maxillofac Implants. 2011;26(4):860–5.
28. Larson BE, Vaubel CJ, Grünheid T. Effectiveness of computer-assisted orthodontic treatment technology to achieve predicted outcomes. Angle Orthod. 2013;83(4):557–62.
29. Li H, Song L, Sun J, Ma J, Shen Z. Dental ceramic prostheses by stereolithography-based additive manufacturing: potentials and challenges. Adv Appl Ceram. 2019;118(1–2):30–6.
30. Tahayeri A, Morgan M, Fugolin AP, Bompolaki D, Athirasala A, Pfeifer CS, Ferracane JL, Bertassoni LE. 3D printed versus conventionally cured provisional crown and bridge dental materials. Dent Mater. 2018;34(2):192–200.
31. Kruth J-P, Wang X, Laoui T, Froyen L. Lasers and materials in selective laser sintering. Assem Autom. 2003;23:357–71.
32. Kruth J-P, Mercelis P, Vaerenbergh JV, Froyen L, Rombouts M. Binding mechanisms in selective laser sintering and selective laser melting. Rapid Prototyp J. 2005;11(1):26–36.
33. Tolochko NK, Khlopkov YV, Mozzharov SE, Ignatiev MB, Laoui T, Titov VI. Absorptance of powder materials suitable for laser sintering. Rapid Prototyp J. 2000;6:155–61.
34. Wang R-J, Wang L, Zhao L, Liu Z. Influence of process parameters on part shrinkage in SLS. Int J Adv Manuf Technol. 2007;33(5–6):498–504.
35. Kruth J-P, Levy G, Klocke F, Childs T. Consolidation phenomena in laser and powder-bed based layered manufacturing. CIRP Ann. 2007;56(2):730–59.
36. Duan B, Cheung WL, Wang M. Optimized fabrication of ca–P/PHBV nanocomposite scaffolds via selective laser sintering for bone tissue engineering. Biofabrication. 2011;3(1):015001.
37. Gibson I, Shi D. Material properties and fabrication parameters in selective laser sintering process. Rapid Prototyp J. 1997;3(4):129–36.
38. Silva DN, De Oliveira MG, Meurer E, Meurer MI, Da Silva JVL, Santa-Bárbara A. Dimensional error in selective laser sintering and 3D-printing of models for craniomaxillary anatomy reconstruction. J Cranio-Maxillofac Surg. 2008;36(8):443–9.

39. Ibrahim D, Broilo TL, Heitz C, de Oliveira MG, de Oliveira HW, Nobre SMW, dos Santos Filho JHG, Silva DN. Dimensional error of selective laser sintering, three-dimensional printing and PolyJet™ models in the reproduction of mandibular anatomy. J Cranio-Maxillofac Surg. 2009;37(3):167–73.

40. Chen J, Zhang Z, Chen X, Zhang C, Zhang G, Xu Z. Design and manufacture of customized dental implants by using reverse engineering and selective laser melting technology. J Prosthet Dent. 2014;112(5):1088–1095.e1.

41. Wiria F, Leong K, Chua C, Liu Y. Poly-ε-caprolactone/hydroxyapatite for tissue engineering scaffold fabrication via selective laser sintering. Acta Biomater. 2007;3(1):1–12.

42. Duan B, Wang M, Zhou WY, Cheung WL, Li ZY, Lu WW. Three-dimensional nanocomposite scaffolds fabricated via selective laser sintering for bone tissue engineering. Acta Biomater. 2010;6(12):4495–505.

43. Mazzoli A, Ferretti C, Gigante A, Salvolini E, Mattioli-Belmonte M. Selective laser sintering manufacturing of polycaprolactone bone scaffolds for applications in bone tissue engineering. Rapid Prototyp J. 2015;21:386–92.

44. Durgun I, Ertan R. Experimental investigation of FDM process for improvement of mechanical properties and production cost. Rapid Prototyp J. 2014;20(3):228–35.

45. Mohan N, Senthil P, Vinodh S, Jayanth N. A review on composite materials and process parameters optimisation for the fused deposition modelling process. Virtual Phys Prototyping. 2017;12(1):47–59.

46. Ligon SC, Liska R, Stampfl J, Gurr M, Mülhaupt R. Polymers for 3D printing and customized additive manufacturing. Chem Rev. 2017;117(15):10212–90.

47. Anitha R, Arunachalam S, Radhakrishnan P. Critical parameters influencing the quality of prototypes in fused deposition modelling. J Mater Process Technol. 2001;118(1):385–8.

48. Mohamed OA, Masood SH, Bhowmik JL. Optimization of fused deposition modeling process parameters: a review of current research and future prospects. Adv Manuf. 2015;3(1):42–53.

49. Kouhi E, Masood S, Morsi Y. Design and fabrication of reconstructive mandibular models using fused deposition modeling. Assem Autom. 2008;28(3):246–54.

50. Maschio F, Pandya M, Olszewski R. Experimental validation of plastic mandible models produced by a "Low-cost" 3-dimensional fused deposition modeling printer. Med Sci Monit. 2016;22:943–57.

51. Meier PJ, Jadhav C, Morrison DA, Hayes AJ, Cunneen S. A novel individual reconstruction of a medial orbital wall blow-out fracture using a bone graft molded intraoperatively using a 3-D model: a case report. Oral Maxillofac Surg Cases. 2016;2(2):19–21.

52. Gronet PM, Waskewicz GA, Richardson C. Preformed acrylic cranial implants using fused deposition modeling: a clinical report. J Prosthet Dent. 2003;90(5):429–33.

53. Espalin D, Arcaute K, Rodriguez D, Medina F, Posner M, Wicker R. Fused deposition modeling of patient-specific polymethylmethacrylate implants. Rapid Prototyp J. 2010;16(3):164–73.

54. Sohmura T, Kusumoto N, Otani T, Yamada S, Wakabayashi K, Yatani H. CAD/CAM fabrication and clinical application of surgical template and bone model in oral implant surgery. Clin Oral Implants Res. 2009;20(1):87–93.

55. Nikzad S, Azari A, Ghassemzadeh A. Modified flapless dental implant surgery for planning treatment in a maxilla including sinus lift augmentation through use of virtual surgical planning and a 3-dimensional model. J Oral Maxillofac Surg. 2010;68(9):2291–8.

56. Subburaj K, Nair C, Rajesh S, Meshram SM, Ravi B. Rapid development of auricular prosthesis using CAD and rapid prototyping technologies. Int J Oral Maxillofac Surg. 2007;36(10):938–43.

57. Edgar J, Tint S. Additive manufacturing technologies: 3D printing, rapid prototyping, and direct digital manufacturing. Johnson Matthey Technol Rev. 2015;59(3):193–8.

58. Magdassi S, Kamyshny A. Nanomaterials for 2D and 3D printing. Wiley; 2017.

59. Kent NJ, Jolivet L, O'Neill P, Brabazon D. An evaluation of components manufactured from a range of materials, fabricated using PolyJet technology. Adv Mater Process Technol. 2017;3(3):318–29.

60. Kumar K, Kumar GS. An experimental and theoretical investigation of surface roughness of poly-jet printed parts: this paper explains how local surface orientation affects surface roughness in a poly-jet process. Virtual Phys Prototyping. 2015;10(1):23–34.
61. Hazeveld A, Slater JJH, Ren Y. Accuracy and reproducibility of dental replica models reconstructed by different rapid prototyping techniques. Am J Orthod Dentofac Orthop. 2014;145(1):108–15.
62. Lee K-Y, Cho J-W, Chang N-Y, Chae J-M, Kang K-H, Kim S-C, Cho J-H. Accuracy of three-dimensional printing for manufacturing replica teeth. Kor J Orthodont. 2015;45(5):217–25.
63. Kim S-Y, Shin Y-S, Jung H-D, Hwang C-J, Baik H-S, Cha J-Y. Precision and trueness of dental models manufactured with different 3-dimensional printing techniques. Am J Orthod Dentofac Orthop. 2018;153(1):144–53.
64. Mai H-N, Lee K-B, Lee D-H. Fit of interim crowns fabricated using photopolymer-jetting 3D printing. J Prosthet Dent. 2017;118(2):208–15.
65. Lewis JA, Gratson GM. Direct writing in three dimensions. Mater Today. 2004;7(7–8):32–9.
66. Calvert P. Inkjet printing for materials and devices. Chem Mater. 2001;13(10):3299–305.
67. De Gans BJ, Duineveld PC, Schubert US. Inkjet printing of polymers: state of the art and future developments. Adv Mater. 2004;16(3):203–13.
68. Tekin E, de Gans B-J, Schubert US. Ink-jet printing of polymers–from single dots to thin film libraries. J Mater Chem. 2004;14(17):2627–32.
69. Malinauskas M, Farsari M, Piskarskas A, Juodkazis S. Ultrafast laser nanostructuring of photopolymers: a decade of advances. Phys Rep. 2013;533(1):1–31.
70. Mostafaei A, Behnamian Y, Krimer YL, Stevens EL, Luo JL, Chmielus M. Effect of solutionizing and aging on the microstructure and mechanical properties of powder bed binder jet printed nickel-based superalloy 625. Mater Des. 2016;111:482–91.
71. Brunello G, Sivolella S, Meneghello R, Ferroni L, Gardin C, Piattelli A, Zavan B, Bressan E. Powder-based 3D printing for bone tissue engineering. Biotechnol Adv. 2016;34(5):740–53.
72. Butscher A, Bohner M, Roth C, Ernstberger A, Heuberger R, Doebelin N, Von Rohr PR, Müller R. Printability of calcium phosphate powders for three-dimensional printing of tissue engineering scaffolds. Acta Biomater. 2012;8(1):373–85.
73. Klammert U, Gbureck U, Vorndran E, Rödiger J, Meyer-Marcotty P, Kübler AC. 3D powder printed calcium phosphate implants for reconstruction of cranial and maxillofacial defects. J Cranio-Maxillofac Surg. 2010;38(8):565–70.
74. Inzana JA, Olvera D, Fuller SM, Kelly JP, Graeve OA, Schwarz EM, Kates SL, Awad HA. 3D printing of composite calcium phosphate and collagen scaffolds for bone regeneration. Biomaterials. 2014;35(13):4026–34.
75. Grant GT, Aita-Holmes C, Liacouras P, Garnes J, Wilson WO Jr. Digital capture, design, and manufacturing of a facial prosthesis: clinical report on a pediatric patient. J Prosthet Dent. 2015;114(1):138–41.
76. Kincade C, McHutchion L, Wolfaardt J. Digital design of patient-specific abutments for the retention of implant-retained facial prostheses. J Prosthet Dent. 2018;120(2):309–12.
77. Mostafaei A, Stevens EL, Ference JJ, Schmidt DE, Chmielus M. Binder jetting of a complex-shaped metal partial denture framework. Addit Manuf. 2018;21:63–8.
78. Naitoh M, Kubota Y, Katsumata A, Ohsaki C, Ariji E. Dimensional accuracy of a binder jet model produced from computerized tomography data for dental implants. J Oral Implantol. 2006;32(6):273–6.
79. Do AV, Khorsand B, Geary SM, Salem AK. 3D printing of scaffolds for tissue regeneration applications. Adv Healthc Mater. 2015;4(12):1742–62.
80. Dutta B, Froes FHS. The additive manufacturing (AM) of titanium alloys. Titanium powder metallurgy: Elsevier; 2015. p. 447–68.
81. Sun J, Zhang FQ. The application of rapid prototyping in prosthodontics. J Prosthodont Implant Esthetic Reconstruct Dent. 2012;21(8):641–4.
82. Yang Y, Chen Y, Wei Y, Li Y. 3D printing of shape memory polymer for functional part fabrication. Int J Adv Manuf Technol. 2016;84(9–12):2079–95.

83. Kong D, Ni X, Dong C, Lei X, Zhang L, Man C, Yao J, Cheng X, Li X. Bio-functional and anti-corrosive 3D printing 316L stainless steel fabricated by selective laser melting. Mater Des. 2018;152:88–101.

84. Dutta Roy T, Simon JL, Ricci JL, Rekow ED, Thompson VP, Parsons JR. Performance of hydroxyapatite bone repair scaffolds created via three-dimensional fabrication techniques. J Biomed Mater Res Part A. 2003;67(4):1228–37.

85. Vorndran E, Klarner M, Klammert U, Grover LM, Patel S, Barralet JE, Gbureck U. 3D powder printing of β-tricalcium phosphate ceramics using different strategies. Adv Eng Mater. 2008;10(12):B67–71.

86. Serra T, Planell JA, Navarro M. High-resolution PLA-based composite scaffolds via 3-D printing technology. Acta Biomater. 2013;9(3):5521–30.

87. Lee J-S, Hong JM, Jung JW, Shim J-H, Oh J-H, Cho D-W. 3D printing of composite tissue with complex shape applied to ear regeneration. Biofabrication. 2014;6(2):024103.

88. Guo T, Holzberg TR, Lim CG, Gao F, Gargava A, Trachtenberg JE, Mikos AG, Fisher JP. 3D printing PLGA: a quantitative examination of the effects of polymer composition and printing parameters on print resolution. Biofabrication. 2017;9(2):024101.

89. Schmidt M, Pohle D, Rechtenwald T. Selective laser sintering of PEEK. CIRP Ann. 2007;56(1):205–8.

90. Rosenzweig D, Carelli E, Steffen T, Jarzem P, Haglund L. 3D-printed ABS and PLA scaffolds for cartilage and nucleus pulposus tissue regeneration. Int J Mol Sci. 2015;16(7):15118–35.

91. Tellis B, Szivek J, Bliss C, Margolis D, Vaidyanathan R, Calvert P. Trabecular scaffolds created using micro CT guided fused deposition modeling. Mater Sci Eng C. 2008;28(1):171–8.

92. Heo EY, Ko NR, Bae MS, Lee SJ, Choi B-J, Kim JH, Kim HK, Park SA, Kwon IK. Novel 3D printed alginate–BFP1 hybrid scaffolds for enhanced bone regeneration. J Ind Eng Chem. 2017;45:61–7.

93. Lewis PL, Green RM, Shah RN. 3D-printed gelatin scaffolds of differing pore geometry modulate hepatocyte function and gene expression. Acta Biomater. 2018;69:63–70.

94. Billiet T, Gevaert E, De Schryver T, Cornelissen M, Dubruel P. The 3D printing of gelatin methacrylamide cell-laden tissue-engineered constructs with high cell viability. Biomaterials. 2014;35(1):49–62.

95. Ouyang L, Highley CB, Rodell CB, Sun W, Burdick JA. 3D printing of shear-thinning hyaluronic acid hydrogels with secondary cross-linking. ACS Biomater Sci Eng. 2016;2(10):1743–51.

96. Liu C, Xia Z, Han Z, Hulley P, Triffitt J, Czernuszka J. Novel 3D collagen scaffolds fabricated by indirect printing technique for tissue engineering. J Biomed Mater Res Part B Appl Biomater. 2008;85(2):519–28.

97. Ghosh S, Parker ST, Wang X, Kaplan DL, Lewis JA. Direct-write assembly of microperiodic silk fibroin scaffolds for tissue engineering applications. Adv Funct Mater. 2008;18(13):1883–9.

98. Murphy SV, Skardal A, Atala A. Evaluation of hydrogels for bio-printing applications. J Biomed Mater Res A. 2013;101(1):272–84.

99. Pati F, Jang J, Ha D-H, Kim SW, Rhie J-W, Shim J-H, Kim D-H, Cho D-W. Printing three-dimensional tissue analogues with decellularized extracellular matrix bioink. Nat Commun. 2014;5:3935.

100. Snyder J, Hamid Q, Wang C, Chang R, Emami K, Wu H, Sun W. Bioprinting cell-laden matrigel for radioprotection study of liver by pro-drug conversion in a dual-tissue microfluidic chip. Biofabrication. 2011;3(3):034112.

101. Tayebi L, Rasoulianboroujeni M, Moharamzadeh K, Almela TK, Cui Z, Ye H. 3D-printed membrane for guided tissue regeneration. Mater Sci Eng C. 2018;84:148–58.

102. Pati F, Ha D-H, Jang J, Han HH, Rhie J-W, Cho D-W. Biomimetic 3D tissue printing for soft tissue regeneration. Biomaterials. 2015;62:164–75.

103. Hermawan H, Ramdan D, Djuansjah JR. Metals for biomedical applications. Biomed Eng Theory Appl. 2011;2011:411–30.

104. Ribeiro F. 3d printing with metals. Comput Control Eng J. 1998;9(1):31–8.

105. Chou D-T, Wells D, Hong D, Lee B, Kuhn H, Kumta PN. Novel processing of iron–manganese alloy-based biomaterials by inkjet 3-D printing. Acta Biomater. 2013;9(10):8593–603.
106. Maleksaeedi S, Wang JK, El-Hajje A, Harb L, Guneta V, He Z, Wiria FE, Choong C, Ruys AJ. Toward 3D printed bioactive titanium scaffolds with bimodal pore size distribution for bone ingrowth. Proc Cirp. 2013;5:158–63.
107. Chevalier J, Gremillard L. Ceramics for medical applications: a picture for the next 20 years. J Eur Ceram Soc. 2009;29(7):1245–55.
108. Kim H-M. Ceramic bioactivity and related biomimetic strategy. Curr Opinion Solid State Mater Sci. 2003;7(4–5):289–99.
109. Yang S, Yang H, Chi X, Evans JR, Thompson I, Cook RJ, Robinson P. Rapid prototyping of ceramic lattices for hard tissue scaffolds. Mater Des. 2008;29(9):1802–9.
110. Will J, Melcher R, Treul C, Travitzky N, Kneser U, Polykandriotis E, Horch R, Greil P. Porous ceramic bone scaffolds for vascularized bone tissue regeneration. J Mater Sci Mater Med. 2008;19(8):2781–90.
111. Fahimipour F, Rasoulianboroujeni M, Dashtimoghadam E, Khoshroo K, Tahriri M, Bastami F, Lobner D, Tayebi L. 3D printed TCP-based scaffold incorporating VEGF-loaded PLGA microspheres for craniofacial tissue engineering. Dent Mater. 2017;33(11):1205–16.
112. Seitz H, Rieder W, Irsen S, Leukers B, Tille C. Three-dimensional printing of porous ceramic scaffolds for bone tissue engineering. J Biomed Mater Res Part B Appl Biomater. 2005;74(2):782–8.
113. Ahn ES. Tricalcium phosphates, their composites, implants incorporating them, and method for their production. Google Patents. 2011.
114. Tarafder S, Dernell WS, Bandyopadhyay A, Bose S. SrO-and MgO-doped microwave sintered 3D printed tricalcium phosphate scaffolds: mechanical properties and in vivo osteogenesis in a rabbit model. J Biomed Mater Res B Appl Biomater. 2015;103(3):679–90.
115. Cox SC, Thornby JA, Gibbons GJ, Williams MA, Mallick KK. 3D printing of porous hydroxyapatite scaffolds intended for use in bone tissue engineering applications. Mater Sci Eng C. 2015;47:237–47.
116. Leukers B, Gülkan H, Irsen SH, Milz S, Tille C, Schieker M, Seitz H. Hydroxyapatite scaffolds for bone tissue engineering made by 3D printing. J Mater Sci Mater Med. 2005;16(12):1121–4.
117. Michna S, Wu W, Lewis JA. Concentrated hydroxyapatite inks for direct-write assembly of 3-D periodic scaffolds. Biomaterials. 2005;26(28):5632–9.
118. Bergmann C, Lindner M, Zhang W, Koczur K, Kirsten A, Telle R, Fischer H. 3D printing of bone substitute implants using calcium phosphate and bioactive glasses. J Eur Ceram Soc. 2010;30(12):2563–7.
119. LeGeros RZ. Properties of osteoconductive biomaterials: calcium phosphates. Clin Orthop Relat Res. 1976-2007;395(2002):81–98.
120. Shanjani Y, De Croos JA, Pilliar RM, Kandel RA, Toyserkani E. Solid freeform fabrication and characterization of porous calcium polyphosphate structures for tissue engineering purposes. J Biomed Mater Res B Appl Biomater. 2010;93(2):510–9.
121. Tarafder S, Bose S. Polycaprolactone-coated 3D printed tricalcium phosphate scaffolds for bone tissue engineering: in vitro alendronate release behavior and local delivery effect on in vivo osteogenesis. ACS Appl Mater Interfaces. 2014;6(13):9955–65.
122. Dávila J, Freitas MSD, Neto PI, Silveira ZDC, Silva JVLD, d'Ávila MA. Fabrication of PCL/β-TCP scaffolds by 3D mini-screw extrusion printing. J Appl Polym Sci. 2016;133(15):43031.
123. Castilho M, Rodrigues J, Pires I, Gouveia B, Pereira M, Moseke C, Groll J, Ewald A, Vorndran E. Fabrication of individual alginate-TCP scaffolds for bone tissue engineering by means of powder printing. Biofabrication. 2015;7(1):015004.
124. Bian W, Li D, Lian Q, Li X, Zhang W, Wang K, Jin Z. Fabrication of a bio-inspired beta-Tricalcium phosphate/collagen scaffold based on ceramic stereolithography and gel casting for osteochondral tissue engineering. Rapid Prototyp J. 2012;18(1):68–80.
125. Wei Q, Deng N-N, Guo J, Deng J. Synthetic polymers for biomedical applications. Int J Biomater. 2018;2018:7158621.

126. Stansbury JW, Idacavage MJ. 3D printing with polymers: challenges among expanding options and opportunities. Dent Mater. 2016;32(1):54–64.
127. Guvendiren M, Molde J, Soares RM, Kohn J. Designing biomaterials for 3D printing. ACS Biomater Sci Eng. 2016;2(10):1679–93.
128. Suganuma J, Alexander H. Biological response of intramedullary bone to poly-L-lactic acid. J Appl Biomater. 1993;4(1):13–27.
129. Woodruff MA, Hutmacher DW. The return of a forgotten polymer—Polycaprolactone in the 21st century. Prog Polym Sci. 2010;35(10):1217–56.
130. Shor L, Güçeri S, Chang R, Gordon J, Kang Q, Hartsock L, An Y, Sun W. Precision extruding deposition (PED) fabrication of polycaprolactone (PCL) scaffolds for bone tissue engineering. Biofabrication. 2009;1(1):015003.
131. Haq RHA, Wahab B, Saidin M, Jaimi NI. Fabrication process of polymer nano-composite filament for fused deposition modeling. Applied Mechanics and Materials: Trans Tech Publ; 2014. p. 8–12.
132. Williams JM, Adewunmi A, Schek RM, Flanagan CL, Krebsbach PH, Feinberg SE, Hollister SJ, Das S. Bone tissue engineering using polycaprolactone scaffolds fabricated via selective laser sintering. Biomaterials. 2005;26(23):4817–27.
133. Yeong W, Sudarmadji N, Yu H, Chua C, Leong K, Venkatraman S, Boey Y, Tan L. Porous polycaprolactone scaffold for cardiac tissue engineering fabricated by selective laser sintering. Acta Biomater. 2010;6(6):2028–34.
134. Temple JP, Hutton DL, Hung BP, Huri PY, Cook CA, Kondragunta R, Jia X, Grayson WL. Engineering anatomically shaped vascularized bone grafts with hASCs and 3D-printed PCL scaffolds. J Biomed Mater Res A. 2014;102(12):4317–25.
135. Bose S, Vahabzadeh S, Bandyopadhyay A. Bone tissue engineering using 3D printing. Mater Today. 2013;16(12):496–504.
136. Konopnicki S, Sharaf B, Resnick C, Patenaude A, Pogal-Sussman T, Hwang K-G, Abukawa H, Troulis MJ. Tissue-engineered bone with 3-dimensionally printed β-tricalcium phosphate and polycaprolactone scaffolds and early implantation: An in vivo pilot study in a porcine mandible model. J Oral Maxillofac Surg. 2015;73(5):1016.e1–1016.e11.
137. Zhu Y, Gao C, Shen J. Surface modification of polycaprolactone with poly (methacrylic acid) and gelatin covalent immobilization for promoting its cytocompatibility. Biomaterials. 2002;23(24):4889–95.
138. Causa F, Battista E, Della Moglie R, Guarnieri D, Iannone M, Netti PA. Surface investigation on biomimetic materials to control cell adhesion: the case of RGD conjugation on PCL. Langmuir. 2010;26(12):9875–84.
139. Chen M, Le DQ, Baatrup A, Nygaard JV, Hein S, Bjerre L, Kassem M, Zou X, Bünger C. Self-assembled composite matrix in a hierarchical 3-D scaffold for bone tissue engineering. Acta Biomater. 2011;7(5):2244–55.
140. Bauermeister AJ, Zuriarrain A, Newman MI. Three-dimensional printing in plastic and reconstructive surgery: a systematic review. Ann Plast Surg. 2016;77(5):569–76.
141. Makadia HK, Siegel SJ. Poly lactic-co-glycolic acid (PLGA) as biodegradable controlled drug delivery carrier. Polymers. 2011;3(3):1377–97.
142. Park TG. Degradation of poly (lactic-co-glycolic acid) microspheres: effect of copolymer composition. Biomaterials. 1995;16(15):1123–30.
143. Wang X, Jiang M, Zhou Z, Gou J, Hui D. 3D printing of polymer matrix composites: a review and prospective. Compos Part B. 2017;110:442–58.
144. Vozzi G, Flaim C, Ahluwalia A, Bhatia S. Fabrication of PLGA scaffolds using soft lithography and microsyringe deposition. Biomaterials. 2003;24(14):2533–40.
145. Ge Z, Tian X, Heng BC, Fan V, Yeo JF, Cao T. Histological evaluation of osteogenesis of 3D-printed poly-lactic-co-glycolic acid (PLGA) scaffolds in a rabbit model. Biomed Mater. 2009;4(2):021001.
146. Mironov AV, Grigoryev AM, Krotova LI, Skaletsky NN, Popov VK, Sevastianov VI. 3D printing of PLGA scaffolds for tissue engineering. J Biomed Mater Res A. 2017;105(1):104–9.

147. Intra J, Glasgow JM, Mai HQ, Salem AK. Pulsatile release of biomolecules from polydimethylsiloxane (PDMS) chips with hydrolytically degradable seals. J Control Release. 2008;127(3):280–7.
148. Mountziaris PM, Spicer PP, Kasper FK, Mikos AG. Harnessing and modulating inflammation in strategies for bone regeneration. Tissue Eng Part B Rev. 2011;17(6):393–402.
149. Rajan V, Murray R. The duplicitous nature of inflammation in wound repair. Wound Pract Res J Aust Wound Manag Assoc. 2008;16(3):122.
150. Talbott MF, Springer GS, Berglund LA. The effects of crystallinity on the mechanical properties of PEEK polymer and graphite fiber reinforced PEEK. J Compos Mater. 1987;21(11):1056–81.
151. Parthasarathy J. 3D modeling, custom implants and its future perspectives in craniofacial surgery. Ann Maxillofac Surg. 2014;4(1):9.
152. Kurtz SM. PEEK biomaterials handbook. William Andrew; 2019.
153. Hoskins T, Dearn K, Kukureka S. Mechanical performance of PEEK produced by additive manufacturing. Polym Test. 2018;70:511–9.
154. El Halabi F, Rodriguez J, Rebolledo L, Hurtós E, Doblaré M. Mechanical characterization and numerical simulation of polyether–ether–ketone (PEEK) cranial implants. J Mech Behav Biomed Mater. 2011;4(8):1819–32.
155. Berretta S, Davies R, Shyng Y, Wang Y, Ghita O. Fused deposition modelling of high temperature polymers: exploring CNT PEEK composites. Polym Test. 2017;63:251–62.
156. Rahman KM, Letcher T, Reese R. Mechanical properties of additively manufactured PEEK components using fused filament fabrication, ASME 2015 international mechanical engineering congress and exposition. American Society of Mechanical Engineers Digital Collection; 2016.
157. Kurtz SM. Chemical and radiation stability of PEEK. In: PEEK biomaterials handbook. Elsevier; 2012. p. 75–9.
158. Rodríguez JF, Thomas JP, Renaud JE. Mechanical behavior of acrylonitrile butadiene styrene (ABS) fused deposition materials. Experimental investigation. Rapid Prototyp J. 2001;7(3):148–58.
159. Rodriguez JF, Thomas JP, Renaud JE. Characterization of the mesostructure of fused-deposition acrylonitrile-butadiene-styrene materials. Rapid Prototyp J. 2000;6(3):175–86.
160. Li Y, Shimizu H. Improvement in toughness of poly (l-lactide)(PLLA) through reactive blending with acrylonitrile–butadiene–styrene copolymer (ABS): morphology and properties. Eur Polym J. 2009;45(3):738–46.
161. Dawoud M, Taha I, Ebeid SJ. Mechanical behaviour of ABS: An experimental study using FDM and injection moulding techniques. J Manuf Process. 2016;21:39–45.
162. Szykiedans K, Credo W. Mechanical properties of FDM and SLA low-cost 3-D prints. Proc Eng. 2016;136:257–62.
163. X. Li, R. Cui, L. Sun, K.E. Aifantis, Y. Fan, Q. Feng, F. Cui, F. Watari, 3-D-printed biopolymers for tissue engineering application, international journal of polymer science 2014 (2014).
164. Hutmacher DW. Scaffolds in tissue engineering bone and cartilage. In: The biomaterials: silver jubilee compendium. Elsevier; 2000. p. 175–89.
165. Szivek J, Bliss C, Geffre C, Margolis D, DeYoung D, Ruth J, Schnepp A, Tellis B, Vaidyanathan R. 213 porous polybutylene terephthalate implants allow for bone ingrowth and provide a well-anchored scaffold that can be used to deliver tissue-engineered cartilage. BMJ Publishing Group Limited; 2006.
166. Radder A, Leenders H, Van Blitterswijk C. Bone-bonding behaviour of poly (ethylene oxide)-polybutylene terephthalate copolymer coatings and bulk implants: a comparative study. Biomaterials. 1995;16(7):507–13.
167. Ullah F, Othman MBH, Javed F, Ahmad Z, Akil HM. Classification, processing and application of hydrogels: a review. Mater Sci Eng C. 2015;57:414–33.
168. Buwalda SJ, Boere KW, Dijkstra PJ, Feijen J, Vermonden T, Hennink WE. Hydrogels in a historical perspective: from simple networks to smart materials. J Control Release. 2014;190:254–73.

169. Draget KI. Alginates. Handbook of hydrocolloids: Elsevier; 2009. p. 807–28.
170. Andresen I-L, Skipnes O, Smidsrød O, Ostgaard K, Hemmer PC. Some biological functions of matrix components in benthic algae in relation to their chemistry and the composition of seawater. ACS Publications; 1977.
171. Shi W, He R, Liu Y. 3D printing scaffolds with hydrogel materials for biomedical applications. Eur J BioMed Res. 2015;1(3):3–8.
172. Mørch YA, Qi M, Gundersen POM, Formo K, Lacik I, Skjåk-Bræk G, Oberholzer J, Strand BL. Binding and leakage of barium in alginate microbeads. J Biomed Mater Res A. 2012;100(11):2939–47.
173. Topuz F, Henke A, Richtering W, Groll J. Magnesium ions and alginate do form hydrogels: a rheological study. Soft Matter. 2012;8(18):4877–81.
174. Lee S-J, Rhie J-W, Cho D-W. Development of three-dimensional alginate encapsulated chondrocyte hybrid scaffold using microstereolithography. J Manuf Sci Eng. 2008;130(2):021007.
175. Lee S, Kang T, Rhie J, Cho D. Development of three-dimensional hybrid scaffold using chondrocyte-encapsulated alginate hydroge (S & M 0695). Sens Mater. 2007;19(8):445.
176. Lee S-J, Kang H-W, Park JK, Rhie J-W, Hahn SK, Cho D-W. Application of microstereolithography in the development of three-dimensional cartilage regeneration scaffolds. Biomed Microdevices. 2008;10(2):233–41.
177. Axpe E, Oyen M. Applications of alginate-based bioinks in 3D bioprinting. Int J Mol Sci. 2016;17(12):1976.
178. Xu C, Chai W, Huang Y, Markwald RR. Scaffold-free inkjet printing of three-dimensional zigzag cellular tubes. Biotechnol Bioeng. 2012;109(12):3152–60.
179. Yu Y, Zhang Y, Martin JA, Ozbolat IT. Evaluation of cell viability and functionality in vessel-like bioprintable cell-laden tubular channels. J Biomech Eng. 2013;135(9):091011.
180. Hockaday L, Kang K, Colangelo N, Cheung P, Duan B, Malone E, Wu J, Girardi L, Bonassar L, Lipson H. Rapid 3D printing of anatomically accurate and mechanically heterogeneous aortic valve hydrogel scaffolds. Biofabrication. 2012;4(3):035005.
181. Duan B, Hockaday LA, Kang KH, Butcher JT. 3D bioprinting of heterogeneous aortic valve conduits with alginate/gelatin hydrogels. J Biomed Mater Res A. 2013;101(5):1255–64.
182. Pourchet LJ, Thepot A, Albouy M, Courtial EJ, Boher A, Blum LJ, Marquette CA. Human skin 3D bioprinting using scaffold-free approach. Adv Healthc Mater. 2017;6(4):1601101.
183. Markstedt K, Mantas A, Tournier I, Ávila HM, Hägg D, Gatenholm P. 3D bioprinting human chondrocytes with nanocellulose–alginate bioink for cartilage tissue engineering applications. Biomacromolecules. 2015;16(5):1489–96.
184. Alsberg E, Anderson K, Albeiruti A, Franceschi R, Mooney D. Cell-interactive alginate hydrogels for bone tissue engineering. J Dent Res. 2001;80(11):2025–9.
185. Lee KY, Mooney DJ. Hydrogels for tissue engineering. Chem Rev. 2001;101(7):1869–80.
186. Su K, Wang C. Recent advances in the use of gelatin in biomedical research. Biotechnol Lett. 2015;37(11):2139–45.
187. Elzoghby AO, Samy WM, Elgindy NA. Protein-based nanocarriers as promising drug and gene delivery systems. J Control Release. 2012;161(1):38–49.
188. Wang H, Boerman OC, Sariibrahimoglu K, Li Y, Jansen JA, Leeuwenburgh SC. Comparison of micro-vs. nanostructured colloidal gelatin gels for sustained delivery of osteogenic proteins: bone morphogenetic protein-2 and alkaline phosphatase. Biomaterials. 2012;33(33):8695–703.
189. Cheng NC, Chang HH, Tu YK, Young TH. Efficient transfer of human adipose-derived stem cells by chitosan/gelatin blend films. J Biomed Mater Res B Appl Biomater. 2012;100(5):1369–77.
190. Ruggeri RR, Bressan FF, Siqueira NM, Meirelles F, Frantz N, Watanabe YF, Soares RM, Bos-Mikich A. Derivation and culture of putative parthenogenetic embryonic stem cells in new gelatin substrates modified with galactomannan. Macromol Res. 2014;22(10):1053–8.
191. Tang G, Zhang H, Zhao Y, Zhang Y, Li X, Yuan X. Preparation of PLGA scaffolds with graded pores by using a gelatin-microsphere template as porogen. J Biomater Sci Polym Ed. 2012;23(17):2241–57.

192. Young S, Wong M, Tabata Y, Mikos AG. Gelatin as a delivery vehicle for the controlled release of bioactive molecules. J Control Release. 2005;109(1–3):256–74.

193. Wang X, Yan Y, Pan Y, Xiong Z, Liu H, Cheng J, Liu F, Lin F, Wu R, Zhang R. Generation of three-dimensional hepatocyte/gelatin structures with rapid prototyping system. Tissue Eng. 2006;12(1):83–90.

194. Wu Z, Su X, Xu Y, Kong B, Sun W, Mi S. Bioprinting three-dimensional cell-laden tissue constructs with controllable degradation, Scientific Reports 6 (2016).

195. Laronda MM, Rutz AL, Xiao S, Whelan KA, Duncan FE, Roth EW, Woodruff TK, Shah RN. A bioprosthetic ovary created using 3D printed microporous scaffolds restores ovarian function in sterilized mice. Nat Commun. 2017;8:15261.

196. Irvine SA, Agrawal A, Lee BH, Chua HY, Low KY, Lau BC, Machluf M, Venkatraman S. Printing cell-laden gelatin constructs by free-form fabrication and enzymatic protein cross-linking. Biomed Microdevices. 2015;17(1):16.

197. Bertassoni LE, Cardoso JC, Manoharan V, Cristino AL, Bhise NS, Araujo WA, Zorlutuna P, Vrana NE, Ghaemmaghami AM, Dokmeci MR. Direct-write bioprinting of cell-laden methacrylated gelatin hydrogels. Biofabrication. 2014;6(2):024105.

198. Zhang L-M. Cellulosic associative thickeners. Carbohydr Polym. 2001;45(1):1–10.

199. Utech S, Boccaccini AR. A review of hydrogel-based composites for biomedical applications: enhancement of hydrogel properties by addition of rigid inorganic fillers. J Mater Sci. 2016;51(1):271–310.

200. Blokhuis TJ, Termaat MF, den Boer FC, Patka P, Bakker FC, Henk JTM. Properties of calcium phosphate ceramics in relation to their in vivo behavior. J Trauma Acute Care Surg. 2000;48(1):179.

201. Song W, Tian M, Chen F, Tian Y, Wan C, Yu X. The study on the degradation and mineralization mechanism of ion-doped calcium polyphosphate in vitro. J Biomed Mater Res Part B Appl Biomater. 2009;89(2):430–8.

202. Khanarian NT, Jiang J, Wan LQ, Mow VC, Lu HH. A hydrogel-mineral composite scaffold for osteochondral interface tissue engineering. Tissue Eng A. 2011;18(5–6):533–45.

203. Gobin AS, West JL. Val-ala-pro-gly, an elastin-derived non-integrin ligand: smooth muscle cell adhesion and specificity. J Biomed Mater Res Part A. 2003;67(1):255–9.

204. Dehghani S, Rasoulianboroujeni M, Ghasemi H, Keshel SH, Nozarian Z, Hashemian MN, Zarei-Ghanavati M, Latifi G, Ghaffari R, Cui Z. 3D-printed membrane as an alternative to amniotic membrane for ocular surface/conjunctival defect reconstruction: An in vitro & in vivo study. Biomaterials. 2018;174:95–112.

205. Shi W, Sun M, Hu X, Ren B, Cheng J, Li C, Duan X, Fu X, Zhang J, Chen H. Structurally and functionally optimized silk-fibroin–gelatin scaffold using 3D printing to repair cartilage injury in vitro and in vivo. Adv Mater. 2017;29(29):1701089.

206. Ng WL, Yeong WY, Naing MW. Polyelectrolyte gelatin-chitosan hydrogel optimized for 3D bioprinting in skin tissue engineering. Int J Bioprint. 2016;2(1):53–62.

207. Malafaya PB, Silva GA, Reis RL. Natural–origin polymers as carriers and scaffolds for biomolecules and cell delivery in tissue engineering applications. Adv Drug Deliv Rev. 2007;59(4–5):207–33.

208. Van Den Bulcke AI, Bogdanov B, De Rooze N, Schacht EH, Cornelissen M, Berghmans H. Structural and rheological properties of methacrylamide modified gelatin hydrogels. Biomacromolecules. 2000;1(1):31–8.

209. Koshy ST, Ferrante TC, Lewin SA, Mooney DJ. Injectable, porous, and cell-responsive gelatin cryogels. Biomaterials. 2014;35(8):2477–87.

210. Burdick JA, Prestwich GD. Hyaluronic acid hydrogels for biomedical applications. Adv Mater. 2011;23(12):H41–56.

211. Kogan G, Šoltés L, Stern R, Gemeiner P. Hyaluronic acid: a natural biopolymer with a broad range of biomedical and industrial applications. Biotechnol Lett. 2007;29(1):17–25.

212. Schuurman W, Levett PA, Pot MW, van Weeren PR, Dhert WJ, Hutmacher DW, Melchels FP, Klein TJ, Malda J. Gelatin-methacrylamide hydrogels as potential biomaterials for fabrication of tissue-engineered cartilage constructs. Macromol Biosci. 2013;13(5):551–61.

213. Pescosolido L, Schuurman W, Malda J, Matricardi P, Alhaique F, Coviello T, van Weeren PR, Dhert WJ, Hennink WE, Vermonden T. Hyaluronic acid and dextran-based semi-IPN hydrogels as biomaterials for bioprinting. Biomacromolecules. 2011;12(5):1831–8.
214. Skardal A, Devarasetty M, Kang H-W, Mead I, Bishop C, Shupe T, Lee SJ, Jackson J, Yoo J, Soker S. A hydrogel bioink toolkit for mimicking native tissue biochemical and mechanical properties in bioprinted tissue constructs. Acta Biomater. 2015;25:24–34.
215. Highley CB, Prestwich GD, Burdick JA. Recent advances in hyaluronic acid hydrogels for biomedical applications. Curr Opin Biotechnol. 2016;40:35–40.
216. Kesti M, Müller M, Becher J, Schnabelrauch M, D'Este M, Eglin D, Zenobi-Wong M. A versatile bioink for three-dimensional printing of cellular scaffolds based on thermally and photo-triggered tandem gelation. Acta Biomater. 2015;11:162–72.
217. Müller M, Becher J, Schnabelrauch M, Zenobi-Wong M. Nanostructured Pluronic hydrogels as bioinks for 3D bioprinting. Biofabrication. 2015;7(3):035006.
218. Skardal A, Zhang J, McCoard L, Xu X, Oottamasathien S, Prestwich GD. Photocrosslinkable hyaluronan-gelatin hydrogels for two-step bioprinting. Tissue Eng A. 2010;16(8):2675–85.
219. Skardal A, Zhang J, Prestwich GD. Bioprinting vessel-like constructs using hyaluronan hydrogels crosslinked with tetrahedral polyethylene glycol tetracrylates. Biomaterials. 2010;31(24):6173–81.
220. Skardal A, Zhang J, McCoard L, Oottamasathien S, Prestwich GD. Dynamically crosslinked gold nanoparticle–hyaluronan hydrogels. Adv Mater. 2010;22(42):4736–40.
221. Parenteau-Bareil R, Gauvin R, Berthod F. Collagen-based biomaterials for tissue engineering applications. Materials. 2010;3(3):1863–87.
222. Silvipriya K, Kumar KK, Bhat A, Kumar BD, John A, Lakshmanan P. Collagen: animal sources and biomedical application. J Appl Pharm Sci. 2015;5(3):123–7.
223. Chevallay B, Herbage D. Collagen-based biomaterials as 3D scaffold for cell cultures: applications for tissue engineering and gene therapy. Med Biol Eng Comput. 2000;38(2):211–8.
224. Yang Y-l, Leone LM, Kaufman LJ. Elastic moduli of collagen gels can be predicted from two-dimensional confocal microscopy. Biophys J. 2009;97(7):2051–60.
225. Glowacki J, Mizuno S. Collagen scaffolds for tissue engineering. Biopolym Orig Res Biomol. 2008;89(5):338–44.
226. Loo Y, Lakshmanan A, Ni M, Toh LL, Wang S, Hauser CA. Peptide bioink: self-assembling nanofibrous scaffolds for three-dimensional organotypic cultures. Nano Lett. 2015;15(10):6919–25.
227. Helary C, Bataille I, Abed A, Illoul C, Anglo A, Louedec L, Letourneur D, Meddahi-Pelle A, Giraud-Guille MM. Concentrated collagen hydrogels as dermal substitutes. Biomaterials. 2010;31(3):481–90.
228. Nöth U, Rackwitz L, Heymer A, Weber M, Baumann B, Steinert A, Schütze N, Jakob F, Eulert J. Chondrogenic differentiation of human mesenchymal stem cells in collagen type I hydrogels. J Biomed Mater Res Part A. 2007;83(3):626–35.
229. Park JY, Choi J-C, Shim J-H, Lee J-S, Park H, Kim SW, Doh J, Cho D-W. A comparative study on collagen type I and hyaluronic acid dependent cell behavior for osteochondral tissue bioprinting. Biofabrication. 2014;6(3):035004.
230. Lee W, Lee V, Polio S, Keegan P, Lee JH, Fischer K, Park JK, Yoo SS. On-demand three-dimensional freeform fabrication of multi-layered hydrogel scaffold with fluidic channels. Biotechnol Bioeng. 2010;105(6):1178–86.
231. Lee W, Debasitis JC, Lee VK, Lee J-H, Fischer K, Edminster K, Park J-K, Yoo S-S. Multi-layered culture of human skin fibroblasts and keratinocytes through three-dimensional freeform fabrication. Biomaterials. 2009;30(8):1587–95.
232. Lee W, Pinckney J, Lee V, Lee J-H, Fischer K, Polio S, Park J-K, Yoo S-S. Three-dimensional bioprinting of rat embryonic neural cells. Neuroreport. 2009;20(8):798–803.
233. Zhao L, Lee VK, Yoo S-S, Dai G, Intes X. The integration of 3-D cell printing and meso-scopic fluorescence molecular tomography of vascular constructs within thick hydrogel scaffolds. Biomaterials. 2012;33(21):5325–32.

234. Lee V, Singh G, Trasatti JP, Bjornsson C, Xu X, Tran TN, Yoo S-S, Dai G, Karande P. Design and fabrication of human skin by three-dimensional bioprinting. Tissue Eng Part C Methods. 2013;20(6):473–84.

235. Lee Y-B, Polio S, Lee W, Dai G, Menon L, Carroll RS, Yoo S-S. Bio-printing of collagen and VEGF-releasing fibrin gel scaffolds for neural stem cell culture. Exp Neurol. 2010;223(2):645–52.

236. Skardal A, Mack D, Kapetanovic E, Atala A, Jackson JD, Yoo J, Soker S. Bioprinted amniotic fluid-derived stem cells accelerate healing of large skin wounds. Stem Cells Transl Med. 2012;1(11):792–802.

237. Xu T, Binder KW, Albanna MZ, Dice D, Zhao W, Yoo JJ, Atala A. Hybrid printing of mechanically and biologically improved constructs for cartilage tissue engineering applications. Biofabrication. 2012;5(1):015001.

238. Hinton TJ, Jallerat Q, Palchesko RN, Park JH, Grodzicki MS, Shue H-J, Ramadan MH, Hudson AR, Feinberg AW. Three-dimensional printing of complex biological structures by freeform reversible embedding of suspended hydrogels. Sci Adv. 2015;1(9):e1500758.

239. Shim J-H, Jang K-M, Hahn SK, Park JY, Jung H, Oh K, Park KM, Yeom J, Park SH, Kim SW. Three-dimensional bioprinting of multilayered constructs containing human mesenchymal stromal cells for osteochondral tissue regeneration in the rabbit knee joint. Biofabrication. 2016;8(1):014102.

240. Wu Z, Su X, Xu Y, Kong B, Sun W, Mi S. Bioprinting three-dimensional cell-laden tissue constructs with controllable degradation. Sci Rep. 2016;6:24474.

241. Lee HJ, Kim YB, Ahn SH, Lee JS, Jang CH, Yoon H, Chun W, Kim GH. A new approach for fabricating collagen/ECM-based bioinks using preosteoblasts and human adipose stem cells. Adv Healthc Mater. 2015;4(9):1359–68.

242. Duarte Campos DF, Blaeser A, Korsten A, Neuss S, Jäkel J, Vogt M, Fischer H. The stiffness and structure of three-dimensional printed hydrogels direct the differentiation of mesenchymal stromal cells toward adipogenic and osteogenic lineages. Tissue Eng A. 2014;21(3–4):740–56.

243. Duarte Campos DF, Blaeser A, Buellesbach K, Sen KS, Xun W, Tillmann W, Fischer H. Bioprinting organotypic hydrogels with improved mesenchymal stem cell remodeling and mineralization properties for bone tissue engineering. Adv Healthc Mater. 2016;5(11):1336–45.

244. Sanjana NE, Fuller SB. A fast flexible ink-jet printing method for patterning dissociated neurons in culture. J Neurosci Methods. 2004;136(2):151–63.

245. Rutz AL, Hyland KE, Jakus AE, Burghardt WR, Shah RN. A multimaterial bioink method for 3D printing tunable, cell-compatible hydrogels. Adv Mater. 2015;27(9):1607–14.

246. Alberts B, Johnson A, Lewis J, Raff M, Roberts K, Walter P. The extracellular matrix of animals. In: Molecular biology of the cell. 4th ed. Garland Science; 2002.

247. Miranda-Nieves D, Chaikof EL. Collagen and elastin biomaterials for the fabrication of engineered living tissues. ACS Biomater Sci Eng. 2016;3(5):694–711.

248. Włodarczyk-Biegun MK, del Campo A. 3D bioprinting of structural proteins. Biomaterials. 2017;134:180–201.

249. Suntivich R, Drachuk I, Calabrese R, Kaplan DL, Tsukruk VV. Inkjet printing of silk nest arrays for cell hosting. Biomacromolecules. 2014;15(4):1428–35.

250. DeSimone E, Schacht K, Jungst T, Groll J, Scheibel T. Biofabrication of 3D constructs: fabrication technologies and spider silk proteins as bioinks. Pure Appl Chem. 2015;87(8):737–49.

251. Jose RR, Brown JE, Polido KE, Omenetto FG, Kaplan DL. Polyol-silk bioink formulations as two-part room-temperature curable materials for 3D printing. ACS Biomater Sci Eng. 2015;1(9):780–8.

252. Das S, Pati F, Chameettachal S, Pahwa S, Ray AR, Dhara S, Ghosh S. Enhanced redifferentiation of chondrocytes on microperiodic silk/gelatin scaffolds: toward tailor-made tissue engineering. Biomacromolecules. 2013;14(2):311–21.

253. Das S, Pati F, Choi Y-J, Rijal G, Shim J-H, Kim SW, Ray AR, Cho D-W, Ghosh S. Bioprintable, cell-laden silk fibroin–gelatin hydrogel supporting multilineage differentiation of stem cells for fabrication of three-dimensional tissue constructs. Acta Biomater. 2015;11:233–46.
254. Parker ST, Domachuk P, Amsden J, Bressner J, Lewis JA, Kaplan DL, Omenetto FG. Biocompatible silk printed optical waveguides. Adv Mater. 2009;21(23):2411–5.
255. Schacht K, Jüngst T, Schweinlin M, Ewald A, Groll J, Scheibel T. Biofabrication of cell-loaded 3D spider silk constructs. Angew Chem Int Ed. 2015;54(9):2816–20.
256. Gregory DA, Zhang Y, Smith PJ, Zhao X, Ebbens SJ. Reactive inkjet printing of biocompatible enzyme powered silk micro-rockets. Small. 2016;12(30):4048–55.
257. Ling S, Zhang Q, Kaplan DL, Omenetto F, Buehler MJ, Qin Z. Printing of stretchable silk membranes for strain measurements. Lab Chip. 2016;16(13):2459–66.
258. Tao H, Marelli B, Yang M, An B, Onses MS, Rogers JA, Kaplan DL, Omenetto FG. Inkjet printing of regenerated silk fibroin: from printable forms to printable functions. Adv Mater. 2015;27(29):4273–9.
259. Xu W, Wang X, Yan Y, Zheng W, Xiong Z, Lin F, Wu R, Zhang R. Rapid prototyping three-dimensional cell/gelatin/fibrinogen constructs for medical regeneration. J Bioact Compat Polym. 2007;22(4):363–77.
260. Li S, Xiong Z, Wang X, Yan Y, Liu H, Zhang R. Direct fabrication of a hybrid cell/hydrogel construct by a double-nozzle assembling technology. J Bioact Compat Polym. 2009;24(3):249–65.
261. Zhao Y, Yao R, Ouyang L, Ding H, Zhang T, Zhang K, Cheng S, Sun W. Three-dimensional printing of Hela cells for cervical tumor model in vitro. Biofabrication. 2014;6(3):035001.
262. Dai X, Ma C, Lan Q, Xu T. 3D bioprinted glioma stem cells for brain tumor model and applications of drug susceptibility. Biofabrication. 2016;8(4):045005.
263. Hughes CS, Postovit LM, Lajoie GA. Matrigel: a complex protein mixture required for optimal growth of cell culture. Proteomics. 2010;10(9):1886–90.
264. Crapo PM, Gilbert TW, Badylak SF. An overview of tissue and whole organ decellularization processes. Biomaterials. 2011;32(12):3233–43.
265. https://www.thermofisher.com/us/en/home/life-science/protein-biology/protein-biology-learning-center/protein-biology-resource-library/pierce-protein-methods/carbodiimide-crosslinker-chemistry.html.
266. Grabarek Z, Gergely J. Zero-length crosslinking procedure with the use of active esters. Anal Biochem. 1990;185(1):131–5.
267. Hwang Y-J, Granelli J, Lyubovitsky J. Effects of zero-length and non-zero-length cross-linking reagents on the optical spectral properties and structures of collagen hydrogels. ACS Appl Mater Interfaces. 2011;4(1):261–7.

3D Bioprinting in Oral and Maxillofacial Surgery

3

3.1　Introduction

3D bioprinting, also known as cell-laden printing, is the construction of tissue- and organ-substitutes by the computer-aided printing of human living cells. In this approach, the bio-ink—a collection of cells, biomaterials, and biomolecules—is printed directly in spatially preset points to resemble the hierarchical 3D structure of the natural tissues [1, 2]. Images acquired from medical imaging methods, such as X-ray, CT, and MRI, are used as the blueprints for creating tissue substitutes. 3D bioprinting follows a bottom-up approach, in which self-assembling layers of bio-inks are printed to form tissues and organs (Fig. 3.1) [3].

In general, 3D bioprinting has advantages over tissue engineering methods using cell-free constructs or constructs that will be seeded by cells after fabrication. Direct printing of living cells limits their distribution and makes it possible to control and tailor the geometry of the final constructs. Furthermore, rapid manufacturing of tissue blocks dramatically enhances the scalability of constructing tissue- and organ substitutes. The main limitations of 3D bioprinting include a restricted choice of materials and possible reduction of cell viability upon printing [3].

This chapter presents a brief history of bioprinting followed by a discussion about how bioprinting can be useful in medical research specifically for disease modeling, cancer models, and drug screening. Bioprintable materials that are known as bio-inks are reviewed in this chapter, and biocompatibility, printability, and bio-degradability are deliberated as the essential characteristics for a material to be employed as a bio-ink. The bio-inks are classified and discussed in three groups of hydrogels, interpenetrating networks (IPNs), and nanocomposites. The bioprinting approach in general is described by considering how 3D bioprinting of tissue substitutes is performed in six sequential steps. Inkjet, microextrusion, and laser-assisted bioprinting techniques are introduced as the three main methods of 3D bioprinting followed by a comprehensive comparison of the method. The oral and maxillofacial applications of each of the methods are described in detail. Barriers to using 3D bioprinting for oral and maxillofacial surgery are discussed in a separate

© Springer Nature Switzerland AG 2021
L. Tayebi et al., *3D Printing in Oral & Maxillofacial Surgery*,
https://doi.org/10.1007/978-3-030-77787-6_3

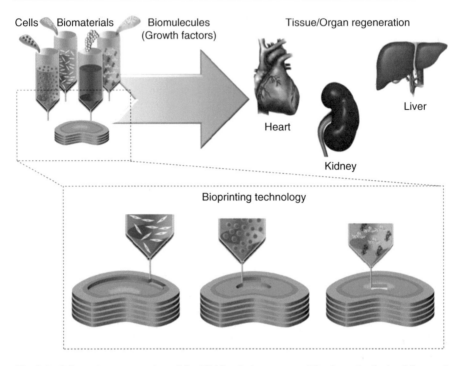

Fig. 3.1 Schematicrepresentation of the 3D bioprinting concept. The figure is obtained from reference [2] with permission

section relying on the drawbacks of the bioprinting and conventional clinical settings. Four main reasons are described for having such barriers including (1) Poor surgical handling of 3D bioprinted constructs, (2) The difficulty of the shipping and storage of 3D bioprinted constructs, and more importantly the preference of using the autologous cells from the patients, which mandate its production inside clinical setting, which are not currently ready to go through the challenging procedure of bioprinting, (3) Limitations in using materials with a specific speed of gelation, viscosity, and cross-linking procedures, and (4) Potential failure of 3D bioprinting to provide a safe environment for cells in long-term. Considering these barriers, this chapter concluded that 3D bioprinting is not a practical approach to be employed in oral and maxillofacial surgery, at least not until the production of suitable patient-specific 3D bioprinted tissue can be translated from laboratory to the practice.

3.2   A Brief History of Bioprinting

The idea of cell and organ printing was originally proposed by Gabor Forgacs, Professor of Bioengineering at the University of Missouri and scientific founder of Organovo and Modern Meadow Inc. In 2003, Forgacs and his team implanted spherical aggregates of Chinese hamster ovary cells into NeuroGel, a biocompatible hydrogel made of poly (N-(2-hydroxypropyl) methacrylamide). Following this scaffold-free cell printing approach, they revealed that the self-assembly of the printed multicellular spheroids allows them to fuse together and form 3D tissue structures. They concluded that the cellular aggregates could be exploited as the building blocks for tissue engineering applications [4]. In the same year, Wilson and Boland invented a jet-based printer for the deposition of molecular and cellular arrays on solid support [5]. In 2004, Barron et al. applied a laser-assisted method to deposit rat cardiac fibroblasts and human sarcoma cells into laser-absorptive Matrigel support. The proposed method made it possible to build stacks of mammalian cells up to 100 μm high for the first time [6].

3D bioprinting of tissue substitutes has progressed through several milestones. In 2010, Binder et al. printed human fibroblasts and keratinocytes onto a full-thickness murine dorsal wound model. The results showed complete epidermis formation and wound closure after 3 weeks. This study revealed for the first time that bioprinting has the potential to repair deep dermal burns and wounds by direct delivery of different skin cell types [7]. In 2013, Duan et al. conducted a study with the goal of building living heart valves. They realized this by seeding aortic valve interstitial cells and aortic root sinus smooth muscle cells within gelatin/alginate hydrogel discs. The key finding of the study was that the tensile properties of the hydrogels were maintained after the incorporation of cells. The bio-printed aortic valve conduits opened up new doors for living cardiac replacements [8]. In another study conducted by Mannoor et al. in 2013, a human ear-shaped cartilaginous construct was fabricated via printing a chondrocyte-laden alginate hydrogel with a coli antenna made from silver nanoparticle-incorporated silicone. Electrical characterizations of the printed bionic ear established its ability to receive and transmit signals beyond audible frequencies [9]. In 2015, Shepherd et al. from Organovo Inc. invented a method to fabricate three-dimensional liver tissue constructs. The bio-printed liver had a multilayered structure composed of parenchymal and non-parenchymal hepatocytes seeded in a gelatin/alginate hydrogel. Each layer of the construct was comprised of one or more liver cell types [10].

3.3 Bioprinting and Medical Research

Successful fabrication of each tissue substitute via 3D bioprinting is a step forward to realize regenerative medical solutions. However, bio-printed constructs have found extra applications in medical research. Presently, most preclinical trials are conducted using animal models; however, there are doubts about the reliability of translating the results of these studies to humans. Furthermore, the production and maintenance of animal models are expensive and laborious. Bio-printed tissue models resemble the composition, structure, and architecture of native tissues and are free of the limitations associated with animal models [11]. Conventional animal models are gradually replaced by three-dimensional bio-printed tissue models. Some of the major applications of bio-printed tissue and organ models include the following:

1. **Disease modeling:** Bio-printed tissue models can help to better understand the mechanisms of diseases. They could also be used for testing therapeutic actions and interventions [12, 13]. With the help of bio-printed tissue models, it is hopeful that regenerative treatment of diseases would be possible in near future.
2. **Cancer models:** Conventional 2D cancer models cannot reflect the natural structure and composition of tumors. Furthermore, monolayer cancer models cannot replicate tumor-stroma interactions. Therefore, they cannot precisely predict the responses of cancer cells to anticancer drugs. By providing detailed information on migration, proliferation, and interactions of cancerous cells, bio-printed cancer models expand our knowledge of tumors and their behaviors. These models could also significantly speed up the discovery and development of anticancer drugs [14, 15].
3. **Drug screening:** 3D tissue models allow the study of dose-dependent responses of tissues to drugs. They also provide a reliable platform to investigate the metabolizing effects and toxic potentials of drugs. Bio-printed tissue models pave the way for personalized drug screening and testing [16, 17].

3.4   Bioprintable Materials: Bio-Inks

Bio-inks play a central role in the bioprinting of tissue and organ substitutes. A critical yet limiting step of bioprinting is the design of the bio-ink. Basically, bio-inks must meet a number of general requirements. Besides, each tissue has its particular biophysical, biochemical, and biomechanical specifications, which should be considered during the formulation of the bio-inks. General and specific requirements determine the design parameters of the bio-inks. To meet the general and tissue-specific requirements, diverse bio-ink formulations with tailored properties are needed [18, 19].

Following are the general requirements for a material to be used as a bio-ink (Table 3.1):

1. **Biocompatibility:** Biocompatibility is the first and the most essential requirement of the bio-inks that should be met before, during, and after the bioprinting

Table 3.1 General requirements of bio-inks [19–21]

Biocompatibility	• Lack of cytotoxicity • Lack of immunogenicity • Ability to modulate cellular adhesion, migration, differentiation, and proliferation
Biodegradability	• Biodegradability • Controllable biodegradation rate
Printability	• Optimal viscosity • Viscoelasticity • Shear-thinning property • Rapid gelation rate

process. Principally, printing living cells requires biocompatible bio-inks. This is necessary to maintain cell survival and function. The bio-inks should also preserve biocompatibility during the printing and post-printing processes. Furthermore, the bio-inks should not cause any harmful local or systemic effects after being implanted within the body. In addition to the bio-inks, products from their in vivo degradation should also be biocompatible. While some authors refer to this requirement as simply being cell compatible, others consider cytocompatibility as the minimum requirement for bio-inks. The latter believe that bio-inks must also have the ability to modulate cellular behaviors including adhesion, migration, differentiation, and proliferation [19, 20].

2. **Biodegradability:** Bio-inks should be composed of biodegradable materials. In vivo degradation of bio-inks allows them to be replaced with the extracellular matrix (ECM) of the printed cells. The biodegradation rate of bio-inks should be finely tunable so that it can be matched to the rate of tissue regeneration and remodeling [20].

3. **Printability:** Printability refers to a set of properties that allows a bio-ink to be deposited layer-by-layer while maintaining its structural integrity and dimensional stability. Printability is the result of a balance between multiple material properties, including viscosity, viscoelasticity, shear-thinning, and gelation kinetics. The bio-ink should have optimal viscosity to obtain uniform cell encapsulation. It should also be viscoelastic and able to protect cells from shear stresses upon the printing process. Furthermore, to avoid mechanical damage to the cells and obtain constructs with high shape fidelity, shear-thinning bio-inks are required. These bio-inks exhibit a non-Newtonian behavior in which their viscosity decreases by increasing the shear rate. The final requirement related to the printability of bio-inks is gelation kinetics, which determines the time needed for the cross-linking of the bio-ink after printing. Bio-inks with rapid gelation are favorable to obtain high print fidelity [19, 21].

Considering the above-mentioned requirements, bio-inks should be designed based on a compromise between biocompatibility and printability. This concept is acknowledged as the "Biofabrication Window". According to this concept, bio-inks with suboptimal print fidelity are utilizable, if they are biocompatible [19].

Classes of bio-inks: Ideal bio-inks are composed of biocompatible components with adjustable in vivo degradation and tissue-resembling physiomechanical properties. In practice, bio-inks are selected from synthetic or naturally derived polymers. Synthetic polymers are of interest because of their high processability, favorable mechanical strength, and the possibility to change properties on demand. However, they lack biomolecular moieties and, thus, cannot modulate cellular behaviors unless they are biofunctionalized. On the other hand, naturally derived polymers facilitate cellular attachment and subsequent migration, proliferation, and differentiation due to their similarity to ECM moieties. Nevertheless, their application is limited by poor mechanical properties and instability at high temperatures and high shear rates [20].

The major classes of bio-inks include hydrogels, interpenetrating networks, and nanocomposites.

1. **Hydrogels:** Hydrogels provide a suitable environment for keeping cells viable and active due to their water retention ability. The highly hydrated medium of hydrogels allows for homogenous cell encapsulation. Furthermore, the three-dimensional structure of hydrogels provides them elasticity and mechanical strength to withstand stresses at the implantation site. Hydrogels are prepared from a wide range of natural and synthetic polymers. Fibrin, collagen, gelatin, alginate, agarose, chitosan, and hyaluronic acid are the most common natural polymers used in hydrogels. These polymers act as biological signals and facilitate cellular proliferation and differentiation. Synthetic polymers mostly used for the preparation of hydrogels include polycaprolactone (PCL), poly(ethylene glycol), polyesters, and pluronics. Although synthetic polymers exhibit favorable printability and processability, they are innately bioinert; therefore, they cannot modulate cellular functions. This limitation could be overcome by strategies that have been developed to biofunctionalize hydrogels. Biochemical signals can be easily embedded within the aqueous medium of hydrogels and improve their bioactivity [20, 22, 23]. Since single component hydrogels do not usually possess both optimal biocompatibility and printability, multimaterial hydrogels are often preferred [19]. For example, it has been shown that alginate exhibits improved cytocompatibility, viscosity, and print fidelity when incorporated by gelatin [24].

2. **Interpenetrating networks (IPNs):** IPNs are composed of multiple different polymer networks that are physically intertwined together. In double network IPNs, the primary network is made from an elastic polymer while the secondary network, which has a lower concentration, is composed of a brittle polymer. The primary and secondary networks provide the flexibility and the stiffness of the IPN, respectively [19]. An example of an IPN is the bio-ink based on ionically crosslinked alginate and covalently crosslinked poly(ethylene glycol) [25].

3. **Nanocomposites:** In nanocomposite bio-inks, nanoparticles are added to hydrogels to improve their bioactivity or enhance their physiomechanical properties. It has been demonstrated that the incorporation of small amounts of nanoparticles into hydrogel networks increases their viscosity, stiffness, and resistance to biodegradation. Furthermore, nanoparticles can provide controlled release of biomolecules and drugs to hydrogels. Several classes of nanoparticles have been used in bio-inks. Among them, clays have attracted increasing interest due to their ubiquity, bioactivity, and unique surface properties. Charged surfaces and high surface/volume ratio of clay particles result in reversible electrostatic interactions with polymer chains. In clay-filled hydrogels, these interactions lead to enhanced viscosity and shear-thinning properties [19]. An example of the application of clays in bio-ink formulations is the study performed by Xavier et al. in 2015. They synthesized a hydrogel by the incorporation of 0.5, 1, and 2% laponite clay into gelatin methacrylate (GelMa). Results indicated the improvement of the bioactivity and mechanical properties of the hydrogel. At high shear

rates, the viscosity of laponite-filled hydrogel was the same as that of the unfilled hydrogel. However, at low shear rates, GelMa-laponite had a greater viscosity. This improved the shear-thinning of the hydrogel and increased the shape fidelity of the constructs. Moreover, higher levels of osteoblast differentiation and mineralized ECM deposition were observed in the nanocomposite hydrogel. This study represented the potential of nanoparticles for improving the processability and printability of bio-inks while providing bioactive signals for encapsulated cells [26].

3.5 Bioprinting Procedural Steps

Bioprinting approaches include biomimicry, autonomous self-assembly, and microtissue-based approaches. Biomimicry, as its name implies, aims to make a replica of the natural components and environment of the target tissue. It exploits scaffolds and bioreactors to mimic the structural and environmental parameters of the natural tissue, respectively. Autonomous self-assembly is based on imitating the embryonic milieu, in which tissue constructs could develop and organize autonomously. In microtissue-based approaches, tissues are fabricated through printing their smallest structural and functional units, called microtissues or mini-tissues. Among the strategies mentioned, scaffolding is required only in biomimicry, and the latter two are scaffold-free [27].

3D bioprinting has significant differences from conventional approaches in tissue engineering. Tissue engineering techniques consist of two sequential steps, including (1) fabrication of a biodegradable scaffold and (2) seeding cells on the fabricated scaffolds. In 3D bioprinting, unlike conventional tissue engineering methods, there is no need to seed cells on scaffolds. Therefore, the problems associated with scaffolds, including degradation and potential immunogenicity of degradation products do not exist in 3D bioprinting. Besides, bioprinting is a single-step process in which cells, biomaterials, and biomolecules are mixed and deposited simultaneously [2, 4].

Bioprinting of tissue substitutes is performed in six consecutive steps (Fig. 3.2):

1. Investigation of the damaged tissue and its surrounding environment using medical imaging techniques.
2. Choosing the design strategy: biomimicry, self-assembly, mini-tissue blocks, or a combination of these.
3. Material selection: polymers (synthetic or natural) or decellularized extracellular matrix (ECM).
4. Selection of differentiated, pluripotent, or multipotent cells from the patient's cells (autogenic cells) or other individuals (allogeneic cells).
5. 3D bioprinting of the tissue substitute via inkjet, microextrusion, or laser-assisted printing.
6. Application of the printed construct in in vitro testing or implantation at the desired site. Maturation in a bioreactor may be required before implantation [21].

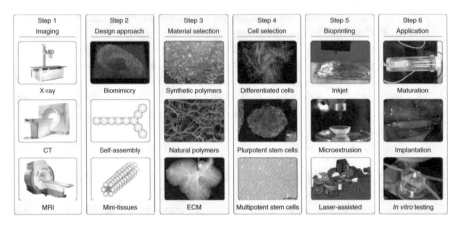

Step 1 Imaging	Step 2 Design approach	Step 3 Material selection	Step 4 Cell selection	Step 5 Bioprinting	Step 6 Application
X-ray	Biomimicry	Synthetic polymers	Differentiated cells	Inkjet	Maturation
CT	Self-assembly	Natural polymers	Pluripotent stem cells	Microextrusion	Implantation
MRI	Mini-tissues	ECM	Multipotent stem cells	Laser-assisted	*In vitro* testing

Fig. 3.2 Steps to bioprinting of tissue substitutes. The picture is obtained from reference [21] with permission

3.6 Common 3D Bioprinting Methods and Their Oral and Maxillofacial Applications

The most common 3D bioprinting methods include inkjet, microextrusion, and laser-assisted bioprinting.

3.6.1 Inkjet Bioprinting

3.6.1.1 Technical Aspects of Inkjet Bioprinting

In inkjet bioprinting (Fig. 3.3), picoliter droplets of bio-ink are deposited into the predefined positions on the substrate. As the substrate moves along the Z axis, the printing process continues in a layer-by-layer order. When the printing process is over, it is needed to apply chemical treatment, pH change, or ultraviolet radiation to solidify the printed liquid [21].

Bio-ink droplets are made using thermal or acoustic energy. In thermal bio-printers, heating the print head creates bubbles that propel the droplets through the nozzle. Heating the print head increases the local temperature up to 300 °C. However, this process lasts only for a few microseconds. Therefore, the overall temperature change is less than 10 °C, and the process does not impair the stability of the bio-molecules or the viability of the cells [28]. Another commonly used energy source for the formation of bio-ink droplets is the acoustic waves generated by piezoelec-tric crystals, which shatter the liquid into droplets at regular intervals [29].

Important characteristics of inkjet bioprinting are summarized in Table 3.2. Low printer cost, high printing speed (up to 10^4 droplets/s), and quick preparation of materials are the advantages of inkjet bioprinting. However, inkjet bioprinting is limited only to biological materials with low viscosity (3–12 mPa.s). Another limi-tation is that the formation of droplets is possible only at low cell concentrations

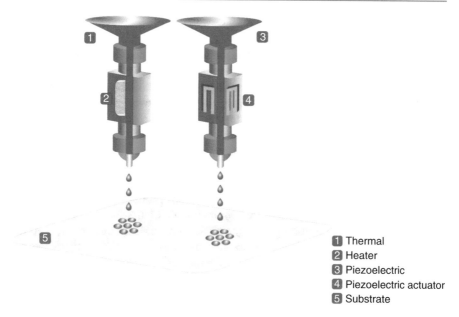

1 Thermal
2 Heater
3 Piezoelectric
4 Piezoelectric actuator
5 Substrate

Fig. 3.3 Inkjet bioprinting modalities

Table 3.2 Characteristics of inkjet bioprinting

Bio-ink viscosity	3–12 mPa.s [30]
Cell density	$<10^6$ cells/mL [3]
Cell viability	>80% [30]
Printing speed	10^4 droplets/s [30]
Resolution	50 μm [31]
Advantages	• Low printer cost • High printing speed • Quick preparation of materials • High resolution
Disadvantages	• Low bio-ink viscosity • Low cell density

($<10^6$ cells/mL). Furthermore, it is almost impossible to achieve uniform droplet size and homogenous cell encapsulation. In addition to all the above, clogging of the nozzle is an inevitable problem that often disrupts the printing process [3, 21, 30].

3.6.1.2 Oral and Maxillofacial Applications of Inkjet Bioprinting

To the best of our knowledge, inkjet bioprinting has not yet been used to fabricate oral and maxillofacial tissue substitutes. However, there are studies that represent the potential application of this method in reconstructing craniofacial defects. In 2012, Cui et al. developed a method for in situ cartilage bioprinting by coupling inkjet bioprinting with an instant photopolymerization apparatus. Immediately after being deposited, the drops of bio-inks—human chondrocytes suspended in poly(ethylene glycol) dimethacrylate (PEGDMA) solution—were irradiated using a UV lamp at 4.5 mW/cm^2 intensity. The method made it possible to print chondrocytes directly at the defect sites with an unprecedented spatial resolution and shape fidelity. Compared to the conventional post-printing polymerization procedure, the instant in situ photopolymerization increased the cell survival rate by 26% [32].

In another study, Gao et al. followed a similar approach to print bone and cartilage substitutes. They printed the bio-ink—bone marrow–derived human mesenchymal stem cells (hMSCs) suspended in peptide-conjugated PEG—using an inkjet bio-printer equipped with a UV lamp. Since the simultaneously printed and polymerized bio-ink had low viscosity, nozzle clogging did not occur during the printing process. The average cell viability at 24 hours after printing was measured to be 87.9%, which indicates excellent cytocompatibility of the bio-ink and minimal shear-induced cell death upon printing. Furthermore, hMSCs encapsulated in the peptide-conjugated PEG hydrogel exhibited enhanced osteogenic and chondrogenic differentiation. More interestingly, cellular hypertrophy was inhibited in the cell-laden peptide-PEG hydrogel [33]. The findings of these studies provide a promising solution for reconstructing craniofacial bone and cartilage defects.

3.6.2   Microextrusion Bioprinting

3.6.2.1 Technical Aspects of Microextrusion Bioprinting

Microextrusion bioprinting is the most common technique of bioprinting. In this method, continuous filaments of the bio-ink extrude through a micro-nozzle and deposit on the substrate. To dispense the bio-ink filaments in the X and Y directions, pneumatic or mechanical pressure is applied (Fig. 3.4). The printed filaments are solidified via chemical treatments, localized heating, or photo-induced crosslinking. By the vertical movement of the nozzle or the substrate, the layers of the microfilaments are arranged on top of each other [21].

Prepolymers, hydrogel solutions, and cell spheroids from a wide range of viscosities (30 mPa.s to 6×10^7 mPa.s) can be printed with microextrusion bioprinting [34]. By printing high viscosity materials, this method has made it feasible to create constructs with physiological cell densities. However, shear stresses exerted to dispense the bio-ink may reduce the viability of the cells [35]. Recently, multi-head extrusion bio-printers have been developed, which use different bio-inks in a single run and facilitate the printing of the multiple tissues [36, 37]. Despite all of the aforesaid benefits, microextrusion bioprinting is a time-consuming method. This is limiting when it comes to printing complex structures with high resolution [21]. Table 3.3 lists the important characteristics of microextrusion bioprinting.

3.6.2.2 Oral and Maxillofacial Applications of Microextrusion Bioprinting

Efforts to fabricate soft and hard tissues through the use of microextrusion bioprinting have just begun in the last decade. In 2008, Fedorovich et al. used a pneumatic

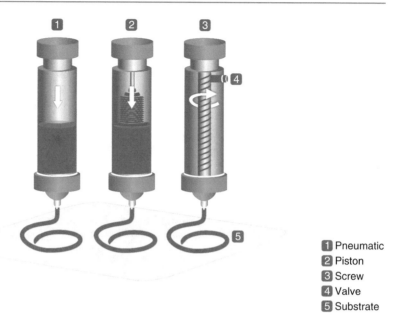

1 Pneumatic
2 Piston
3 Screw
4 Valve
5 Substrate

Fig. 3.4 Microextrusion bioprinting: pneumatic and mechanical (piston and screw) systems

Table 3.3 Characteristics of microextrusion bioprinting

Bio-ink viscosity	30 mPa.s to 6×10^7 mPa.s [34]
Cell density	$>10^8$ cells/mL [21]
Cell viability	40–95% [30]
Printing speed	0.01–0.05 mm/s [30]
Resolution	100 μm [31]
Advantages	High cell density Ability to print materials of a wide range of viscosities
Disadvantages	Low printing speed The possibility of cell damage caused by shear stresses

micro-extruder to dispense alginate hydrogel filaments loaded with bone marrow stromal cells (BMSCs). The four-layered structures were printed at 1–30 mm/s. 0.5–3.0 bar pressure was applied to extrude filaments with 150 μm thickness. The viability of the printed cells was compared to the unprinted cells encapsulated in the alginate hydrogel. The results indicated that the printing process does not significantly affect the survival rate of the cells at 5 hours, 1 day, and 3 days after printing. Furthermore, the alkaline phosphatase activity assay of the cell-laden constructs cultured in the osteogenic medium showed that the printed BMSCs retain their differentiation potential [38].

In 2015, Markstedt et al. designed a bio-ink based on alginate and nano-fibrillated cellulose (NFC) for cartilage bioprinting (Fig. 3.5). They used an extrusion-based bio-printer with a 300-micron diameter nozzle to print the chondrocyte-laden

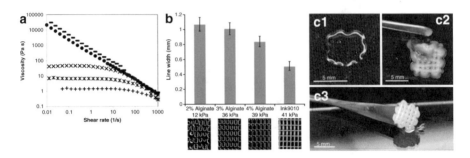

Fig. 3.5 The bio-ink development for bioprinting of cartilaginous structures. Panel A and B present the flow curves and line width using different compositions of NFC and alginate (Ink9010 means NFC/alginate ratio is 90:10). Panels C1, C2, and C3 present the photo of scaffolds made by 3% alginate, 2.5% NFC, and Ink9010, respectively. The figure is obtained from reference [39] with permission. Such bio-inks for cartilage regeneration may have applications in oral and maxillofacial surgery

constructs at 10–20 mm/s. The shear-thinning properties of NFC combined with the potential of alginate to be cross-linked made NFC/alginate a favorable bio-ink to create constructs with high shape fidelity. The high survival rate of the chondrocytes (86% on the seventh day after culture) indicated that the proposed formulation has no cytotoxic effects. The results of the study suggest that nanocellulose-based hydrogel could serve as a promising bio-ink to successfully print cartilaginous structures with anatomical shape [39].

In 2018, Athirasala et al. developed a bio-ink for tooth regeneration. The bio-ink was prepared by mixing 3% (w/v) sodium alginate with dentin matrix isolated from human third molars. During the extrusion of the bio-ink, calcium chloride ($CaCl_2$) solution was pumped through an external syringe to the tip of the nozzle. The contact between $CaCl_2$ and the bio-ink led to the formation of covalently cross-linked fibers. A customized extrusion bio-printer was used to print the bio-ink at 45 μL/min. The authors examined 2:1, 1:1, and 1:2 ratios of alginate to dentin matrix to find the optimal combination with the highest biocompatibility and printability. The alginate contributed to the printability and shape fidelity of the bio-ink, while the demineralized dentin matrix provided soluble and insoluble signals to promote the odontogenic differentiation of the encapsulated human stem cells of the apical papilla (SCAPs). Results showed that the maximum cell survival rate and desirable rheological properties were obtained by using the 1:1 alginate:dentin matrix bio-ink. The authors concluded that the proposed hybrid bio-ink has the potential to be utilized in the regeneration of dental and craniofacial tissues [40].

3.6.3 Laser-Assisted Bioprinting

3.6.3.1 Technical Aspects of Laser-Assisted Bioprinting
In laser-assisted bioprinting, a pulsed laser beam is projected on a scanning mirror. The beam is then transferred to a focusing lens, which irradiates focused pulses on

the ribbon. The ribbon is a glass slide covered by a thin metallic film to absorb the energy of the laser irradiation. The surface of the ribbon is coated by the bio-ink, and the energy absorbed at the surface of the ribbon evaporates the bio-ink and creates droplets. The microdroplets are deposited onto the facing substrate made of glass or quartz (Fig. 3.6). The process is repeated to print the layers of the construct one after another [41].

Table 3.4 summarizes the important characteristics of laser-assisted bioprinting. Laser-assisted bioprinting is a fast printing process that makes it feasible to print up to 1600 mm of the construct per second. This nozzle-free approach allows for the deposition of cells at the density of 10^8 cells/mL with 1 cell/drop resolution. Furthermore, this technique is compatible with a varied range of bio-inks having different viscosities. However, limited construct dimensions, high cost, and time-consuming preparation are the drawbacks of this method. Compared to inkjet and microextrusion-based methods, laser-assisted bioprinting is less common [21, 41].

3.6.3.2 Oral and Maxillofacial Applications of Laser-Assisted Bioprinting

In comparison with inkjet and microextrusion-based methods, laser-assisted bioprinting is less common. In 2011, Catros et al. conducted a pioneering study to create two- and three-dimensional patterns of nano-hydroxyapatite (nHA) and human osteoprogenitor cells via laser-assisted bioprinting. They utilized an infrared Nd:YAG source ($\lambda = 1064$ nm) to emit 30 ns laser pulses. The MG63 cells were printed at the density of 3×10^6 cells/mL. The printed materials were characterized by physicochemical and morphological analysis methods. Furthermore, the viability, adhesion, and proliferation of the printed cells were studied up to 15 days after printing. The results confirmed the biocompatibility of the printed nHA layers. It

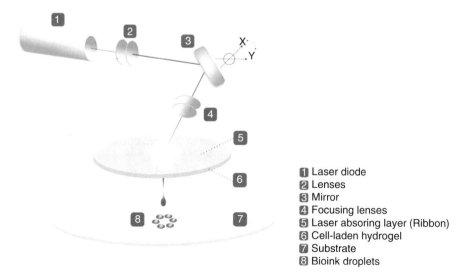

1 Laser diode
2 Lenses
3 Mirror
4 Focusing lenses
5 Laser absoring layer (Ribbon)
6 Cell-laden hydrogel
7 Substrate
8 Bioink droplets

Fig. 3.6 Laser-assisted bioprinting method

Table 3.4 Characteristics of laser-assisted bioprinting

Bio-ink viscosity	1–300 mPa.s [3]
Cell density	10^8 cells/mL [21]
Cell viability	>95% [30]
Printing speed	200–1600 mm/s [30]
Resolution	10–50 μm [30, 42]
Advantages	• High printing resolution • High cell viability
Disadvantages	• High cost • Limited construct dimensions

was also demonstrated that the printing process does not impair the viability or the morphology of the cells. Furthermore, the printed cells preserved their osteoblastic phenotype over the study period [43].

In a similar study, a combination of electrospinning and laser-assisted bioprinting was proposed to fabricate cell-laden bone tissue substitutes [44]. The bio-ink was prepared by adding the suspension of MG63 cells (50×10^6 cells/mL) to 1% (w/v) sodium alginate solution. A 1064 nm Nd:YAG laser was used to generate 18 mJ laser pulses at 5 kHz. The droplets of the bio-ink were deposited on a poly-caprolactone (PCL) electrospun scaffold of 100 μm thickness. The sequential layers of the bio-ink and disk-shaped PCL membranes were printed in a layer-by-layer arrangement. The constructs were either cultured in vitro or implanted in 4-mm wide rat calvarial defects. The results of the cytotoxicity assay revealed that the proposed procedure maintains the in vitro survival of the cells and increases their in vivo proliferation rate. SEM micrographs showed the spreading of the cells on the surface of the PCL membranes. The secretion of the extracellular matrix between PCL fibers was also observed. The histological analyses indicated the formation of dense fibrous tissue in the defects after the eighth week. The authors concluded that combining laser-assisted bioprinting and electrospinning preserves the position and viability of the printed cells and allows for safe handling and implantation of the cell-containing constructs [44].

In 2017, Kerique et al. used laser-assisted bioprinting for in situ bone regeneration in mice calvaria defects as shown in Fig. 3.7 [45]. The ink was prepared by dispersing 1.2% (w/v) of nano-hydroxyapatite (nHA) in 2 mg/mL of type I rat collagen solution. The bio-ink contained 120×10^6 cells/mL of murine bone marrow-derived D1 mesenchymal stromal cells. The three-layered constructs were created in ring and disk shapes by printing a cellularized layer between two layers of collagen/nHA. A 7 W Nd:YAG lamp at 1046 nm was used to generate laser pulses at 30 ns duration. One and 2 months after printing, new bone formation was evaluated using X-ray microtomography (μCT). Also, hematoxylin/eosin/saffron (HES) staining was completed for histological analysis of bone repair in the defects. The results indicated that the disk-shaped constructs promoted homogenous bone formation throughout the defect. However, in the constructs with the ring geometry, marginal bone regeneration was observed at the periphery of the defect. This study presented the promising application of laser-assisted bioprinting for in situ regeneration of

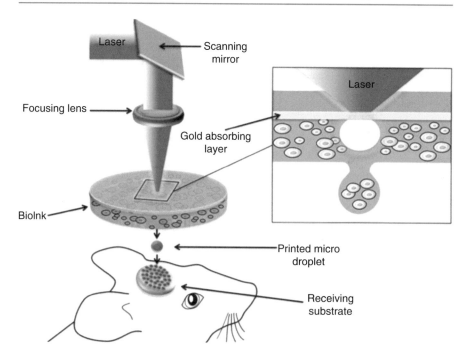

Fig. 3.7 Using laser-assisted bioprinting method to perform in situ printing of mesenchymal stromal cells for bone regeneration. The picture is obtained from reference [45] which is under the terms of the Creative Commons CC BY license and no permission is required for its use

bone defects [45]. This research group has also recently (2020) organized their protocol and published it as a paper entitled "Laser-Assisted Bioprinting for Bone Repair" that offers the detailed protocol to use laser-assisted bioprinting of mesenchymal stromal cells in association with collagen and nanohydroxyapatite, to induce regeneration of bone in a mice calvaria defect model [46].

3.6.4   Comparison of Common Bioprinting Methods

Table 3.5 compares the important characteristics of common bioprinting methods. As the table shows, each method is a compromise of advantages and disadvantages. Inkjet bioprinting allows for rapid printing of low viscosity bio-inks with high resolution. However, the cellular density of the constructs fabricated by inkjet bioprinting is lower than other methods. On the other hand, microextrusion bioprinting is compatible with bio-inks from a wide range of viscosities. However, its printing speed and resolution are below than the other methods. Laser-assisted bioprinting provides superior resolution and cell viability, but it cannot compete with the other two methods in terms of cost.

Table 3.5 Comparison of common 3D bioprinting methods

Properties	Inkjet bioprinting	Microextrusion bioprinting	Laser-assisted bioprinting
Bio-ink viscosity	Low (3–12 mPa.s)	Medium–high ($30^{-6} \times 10^7$ mPa.s)	Low–medium (1–300 mPa.s)
Cell density	Low (<10^6 cells/mL)	High (>10^8 cells/mL)	Medium (10^8 cells/mL)
Cell viability	>80%	40–95%	>95%
Printing speed	Fast (10^4 droplets/s)	Slow (0.01–0.05 mm/s)	Medium (200–1600 mm/s)
Resolution	High (50 μm)	Moderate (100 μm)	High (10–50 μm)
Cost	Low	Low–moderate	High

3.7   Barriers in Using 3D Bioprinting for Oral and Maxillofacial Surgery

A few applications of 3D bioprinting are described in the earlier sections, but comparing these applications with the ones associated with 3D printing, one may consider bioprinting as a less practical approach to be used in oral and maxillofacial surgery. In other words, although 3D bioprinting has drawn the attention of many scientists recently [21, 47], there are some barriers that limit its application as a suitable strategy in oral and maxillofacial surgery. We have summarized these barriers as below:

A. 3D bioprinted constructs are often too soft without secure surgical handling (they collapse when held by hand). The mechanical resistance and stiffness of bioprinted constructs are much lower than the mechanical property values of most living tissues [48].

B. Considering that bioprinted constructs are made of a combination of cells and biomaterials, their storage and shipping are not trivial. On the other hand, we cannot expect the surgeons or their assistants to be able to do the difficult procedure of bioprinting in a conventional clinical setting. It is much practical to ship cell-free patient-specific 3D printed constructs to surgeons prior to surgery. Moreover, if cells are required to be in the construct, it is highly preferable to use autologous cells. In such cases, the clinical settings can be provided with the 3D printed constructs, and then the surgeons can be requested to follow the relevant protocols to seed and incubate the construct with autologous cells. Thus, generally using 3D printed constructs is more practical than 3D bioprinted ones in oral and maxillofacial surgery.

C. The formulations of hydrogels that can be used for bioprinting need to meet specific criteria regarding their speed of gelation, viscosity, and cross-linking, which limit the potential materials that can be used for bioprinting [49].

D. 3D bioprinting may not provide a safe environment for cells in terms of keeping them alive for the long term. There are problems with the growth of the cells in bioprinting constructs such as slow proliferation and possible immobilization within the hydrogels, which hinder their migration [50].

3.8 Summary

3D bioprinting has become a rapidly growing approach in tissue engineering as a promising tool for medical research during the last two decades. Considerable advances made in bioprinting materials and methods have brought hope for regenerative solutions. Among the different methods, inkjet bioprinting, microextrusion, and laser-assisted bioprinting are the most common. Besides, several classes of bioinks have been developed including hydrogels, interpenetrating networks, and nanocomposites. These innovative materials and methods may have the potential in creating tissue substitutes of diverse cell types and biomaterials. Some potential applications in reconstructing hard and soft oromaxillofacial tissues are assumed for 3D bioprinting, which are discussed in this chapter. However, the limitations associated with the nature of the method itself and capabilities of conventional clinical settings, make this method impractical to be used in oral and maxillofacial surgery in near future.

References

1. Mironov V, Boland T, Trusk T, Forgacs G, Markwald RR. Organ printing: computer-aided jet-based 3D tissue engineering. Trends Biotechnol. 2003;21(4):157–61.
2. Seol Y-J, Kang H-W, Lee SJ, Atala A, Yoo JJ. Bioprinting technology and its applications. Eur J Cardiothorac Surg. 2014;46(3):342–8.
3. Mandrycky C, Wang Z, Kim K, Kim D-H. 3D bioprinting for engineering complex tissues. Biotechnol Adv. 2016;34(4):422–34.
4. Jakab K, Neagu A, Mironov V, Markwald RR, Forgacs G. Engineering biological structures of prescribed shape using self-assembling multicellular systems. Proc Natl Acad Sci. 2004;101(9):2864–9.
5. Wilson WC, Boland T. Cell and organ printing 1: protein and cell printers. Anat Rec A: Discov Mol Cell Evol Biol. 2003;272(2):491–6.
6. Barron JA, Ringeisen BR, Kim H, Spargo BJ, Chrisey DB. Application of laser printing to mammalian cells. Thin Solid Films. 2004;453:383–7.
7. Binder KW, Zhao W, Aboushwareb T, Dice D, Atala A, Yoo JJ. In situ bioprinting of the skin for burns. J Am Coll Surg. 2010;211(3):S76.
8. Duan B, Hockaday LA, Kang KH, Butcher JT. 3D bioprinting of heterogeneous aortic valve conduits with alginate/gelatin hydrogels. J Biomed Mater Res A. 2013;101(5):1255–64.
9. Mannoor MS, Jiang Z, James T, Kong YL, Malatesta KA, Soboyejo WO, Verma N, Gracias DH, McAlpine MC. 3D printed bionic ears. Nano Lett. 2013;13(6):2634–9.
10. Shepherd BR, Robbins JB, Gorgen VA, Presnell SC, Engineered liver tissues, arrays thereof, and methods of making the same. U.S. Patent No. 9,222,932. 2015.
11. Ma X, Liu J, Zhu W, Tang M, Lawrence N, Yu C, Gou M, Chen S. 3D bioprinting of functional tissue models for personalized drug screening and in vitro disease modeling. Adv Drug Deliv Rev. 2018;132:235–51.
12. Ma Z, Koo S, Finnegan MA, Loskill P, Huebsch N, Marks NC, Conklin BR, Grigoropoulos CP, Healy KE. Three-dimensional filamentous human diseased cardiac tissue model. Biomaterials. 2014;35(5):1367–77.
13. Bhatia SN, Underhill GH, Zaret KS, Fox IJ. Cell and tissue engineering for liver disease. Sci Transl Med. 2014;6(245):245sr2.
14. McMillin DW, Negri JM, Mitsiades CS. The role of tumour–stromal interactions in modifying drug response: challenges and opportunities. Nat Rev Drug Discov. 2013;12(3):217.

15. Leonard F, Godin B. 3D in vitro model for breast cancer research using magnetic levitation and bioprinting method. In: Breast Cancer. Springer; 2016. p. 239–51.
16. Peng W, Datta P, Ayan B, Ozbolat V, Sosnoski D, Ozbolat IT. 3D bioprinting for drug discovery and development in pharmaceutics. Acta Biomater. 2017;57:26–46.
17. Kizawa H, Nagao E, Shimamura M, Zhang G, Torii H. Scaffold-free 3D bio-printed human liver tissue stably maintains metabolic functions useful for drug discovery. Biochem Biophys Rep. 2017;10:186–91.
18. Malda J, Groll J. A step towards clinical translation of biofabrication. Trends Biotechnol. 2016;34(5):356–7.
19. Chimene D, Lennox KK, Kaunas RR, Gaharwar AK. Advanced bioinks for 3D printing: a materials science perspective. Ann Biomed Eng. 2016;44(6):2090–102.
20. Jose RR, Rodriguez MJ, Dixon TA, Omenetto F, Kaplan DL. Evolution of bioinks and additive manufacturing technologies for 3D bioprinting. ACS Biomater Sci Eng. 2016;2(10):1662–78.
21. Murphy SV, Atala A. 3D bioprinting of tissues and organs. Nat Biotechnol. 2014;32(8):773.
22. Malda J, Visser J, Melchels FP, Jüngst T, Hennink WE, Dhert WJ, Groll J, Hutmacher DW. 25th anniversary article: engineering hydrogels for biofabrication. Adv Mater. 2013;25(36):5011–28.
23. Ji S, Guvendiren M. Recent advances in bioink design for 3D bioprinting of tissues and organs. Front Bioeng Biotechnol. 2017;5:23.
24. Chung JH, Naficy S, Yue Z, Kapsa R, Quigley A, Moulton SE, Wallace GG. Bio-ink properties and printability for extrusion printing living cells. Biomater Sci. 2013;1(7):763–73.
25. Hong S, Sycks D, Chan HF, Lin S, Lopez GP, Guilak F, Leong KW, Zhao X. 3D printing of highly stretchable and tough hydrogels into complex, cellularized structures. Adv Mater. 2015;27(27):4035–40.
26. Xavier JR, Thakur T, Desai P, Jaiswal MK, Sears N, Cosgriff-Hernandez E, Kaunas R, Gaharwar AK. Bioactive nanoengineered hydrogels for bone tissue engineering: a growth-factor-free approach. ACS Nano. 2015;9(3):3109–18.
27. Bishop ES, Mostafa S, Pakvasa M, Luu HH, Lee MJ, Wolf JM, Ameer GA, He T-C, Reid RR. 3-D bioprinting technologies in tissue engineering and regenerative medicine: current and future trends. Genes Dis. 2017;4:185–95.
28. Cui X, Dean D, Ruggeri ZM, Boland T. Cell damage evaluation of thermal inkjet printed Chinese hamster ovary cells. Biotechnol Bioeng. 2010;106(6):963–9.
29. Xu T, Kincaid H, Atala A, Yoo JJ. High-throughput production of single-cell microparticles using an inkjet printing technology. J Manuf Sci Eng. 2008;130(2):021017.
30. Vijayavenkataraman S, Yan W-C, Lu WF, Wang C-H, Fuh JYH. 3D bioprinting of tissues and organs for regenerative medicine. Adv Drug Deliv Rev. 2018;12:296–332.
31. Ozbolat IT, Yu Y. Bioprinting toward organ fabrication: challenges and future trends. IEEE Trans Biomed Eng. 2013;60(3):691–9.
32. Cui X, Breitenkamp K, Finn M, Lotz M, D'Lima DD. Direct human cartilage repair using three-dimensional bioprinting technology. Tissue Eng A. 2012;18(11–12):1304–12.
33. Gao G, Yonezawa T, Hubbell K, Dai G, Cui X. Inkjet-bioprinted acrylated peptides and PEG hydrogel with human mesenchymal stem cells promote robust bone and cartilage formation with minimal printhead clogging. Biotechnol J. 2015;10(10):1568–77.
34. Hölzl K, Lin S, Tytgat L, Van Vlierberghe S, Gu L, Ovsianikov A. Bioink properties before, during and after 3D bioprinting. Biofabrication. 2016;8(3):032002.
35. Panwar A, Tan LP. Current status of bioinks for micro-extrusion-based 3D bioprinting. Molecules. 2016;21(6):685.
36. Kolesky DB, Truby RL, Gladman AS, Busbee TA, Homan KA, Lewis JA. 3D bioprinting of vascularized, heterogeneous cell-laden tissue constructs. Adv Mater. 2014;26(19):3124–30.
37. Lee J-S, Hong JM, Jung JW, Shim J-H, Oh J-H, Cho D-W. 3D printing of composite tissue with complex shape applied to ear regeneration. Biofabrication. 2014;6(2):024103.
38. Fedorovich NE, De Wijn JR, Verbout AJ, Alblas J, Dhert WJ. Three-dimensional fiber deposition of cell-laden, viable, patterned constructs for bone tissue printing. Tissue Eng A. 2008;14(1):127–33.

39. Markstedt K, Mantas A, Tournier I, Ávila HM, Hägg D, Gatenholm P. 3D bioprinting human chondrocytes with nanocellulose–alginate bioink for cartilage tissue engineering applications. Biomacromolecules. 2015;16(5):1489–96.
40. Athirasala A, Tahayeri A, Thrivikraman G, França CM, Monteiro N, Tran V, Ferracane J, Bertassoni LE. A dentin-derived hydrogel bioink for 3D bioprinting of cell laden scaffolds for regenerative dentistry. Biofabrication. 2018;10(2):024101.
41. Guillemot F, Souquet A, Catros S, Guillotin B, Lopez J, Faucon M, Pippenger B, Bareille R, Rémy M, Bellance S. High-throughput laser printing of cells and biomaterials for tissue engineering. Acta Biomater. 2010;6(7):2494–500.
42. Derakhshanfar S, Mbeleck R, Xu K, Zhang X, Zhong W, Xing M. 3D bioprinting for biomedical devices and tissue engineering: a review of recent trends and advances. Bioactive Mater. 2018;3(2):144–56.
43. Catros S, Fricain J-C, Guillotin B, Pippenger B, Bareille R, Remy M, Lebraud E, Desbat B, Amédée J, Guillemot F. Laser-assisted bioprinting for creating on-demand patterns of human osteoprogenitor cells and nano-hydroxyapatite. Biofabrication. 2011;3(2):025001.
44. Catros S, Guillemot F, Nandakumar A, Ziane S, Moroni L, Habibovic P, van Blitterswijk C, Rousseau B, Chassande O, Amédée J. Layer-by-layer tissue microfabrication supports cell proliferation in vitro and in vivo. Tissue Eng Part C Methods. 2011;18(1):62–70.
45. Keriquel V, Oliveira H, Rémy M, Ziane S, Delmond S, Rousseau B, Rey S, Catros S, Amédée J, Guillemot F. In situ printing of mesenchymal stromal cells, by laser-assisted bioprinting, for in vivo bone regeneration applications. Sci Rep. 2017;7(1):1778.
46. Hakobyan D, Kerouredan O, Remy M, Dusserre N, Medina C, Devillard R, Fricain J-C, Oliveira H. Laser-assisted bioprinting for bone repair. In: 3D bioprinting. Springer; 2020. p. 135–44.
47. Mironov V, Reis N, Derby B. Bioprinting: a beginning. Tissue Eng. 2006;12(4):631–4.
48. Schuurman W, Khristov V, Pot MW, van Weeren PR, Dhert WJ, Malda J. Bioprinting of hybrid tissue constructs with tailorable mechanical properties. Biofabrication. 2011;3(2):021001.
49. Skardal A, Atala A. Biomaterials for integration with 3-D bioprinting. Ann Biomed Eng. 2015;43(3):730–46.
50. Mironov V, Visconti RP, Kasyanov V, Forgacs G, Drake CJ, Markwald RR. Organ printing: tissue spheroids as building blocks. Biomaterials. 2009;30(12):2164–74.

3D Printed Medical Modeling for Oral and Maxillofacial Surgeries

4

4.1   Introduction

Technological progress in rapid prototyping and medical imaging have made it possible to create 3D models of different tissues and organs [1]. Medical modeling, also called biomodeling, refers to the direct fabrication of real-sized anatomical models from medical imaging data [2]. The first medical model fabricated via 3D printing was introduced in 1991 at the Clinic for Maxillofacial Surgery in Vienna [3]. Since then, medical models have been increasingly used in dental [4–7], oromaxillofacial [8–12], neurological [13–16], and orthopedic surgeries [17–19]. The most common applications of medical models are for diagnosis, treatment planning, implant design and fabrication, medical education, and communication with patients [20, 21].

Medical models are robust tools to accurately determine the position and dimensions of anatomical defects and deformities. Prior to every surgical intervention, the surgeons need to exactly recognize the morphology of the anomalies and their relationships with the surrounding tissues. This is necessary to find the best method to access the pathological site and perform corrective interventions. 3D anatomical models offer details that are not available via 2D or 3D medical images, nor through the radiologists' reports. Direct visualization of the anatomical structures facilitates diagnostic procedures and allows surgeons to perform simulated surgeries. This improves the precision of the diagnosis and increases the safety of the operation [22–24]. Besides, medical models make it possible to preoperatively design and fabricate patient-specific implants. They also allow for simulating the placement and stabilization of the implants. Therefore, using 3D medical models reduces the operation time by eliminating the need for intraoperative contouring and fixation of the implants [25–27]. Consequently, the treatment outcomes are improved by the use of medical models.

From another point of view, patients and their caregivers need to be provided with necessary information about the state of the disease, the probable diagnosis, and the treatment procedures to be used. Medical models, as physical, graspable

© Springer Nature Switzerland AG 2021
L. Tayebi et al., *3D Printing in Oral & Maxillofacial Surgery*,
https://doi.org/10.1007/978-3-030-77787-6_4

replicas of patients' anatomy, demonstrate the pathological status of the tissues and help patients to better understand their disease. Furthermore, the required treatments and expected outcomes could be presented to the patients with the aid of the medical models. Therefore, medical models improve patient-surgeon communication and facilitate obtaining informed consent from patients and their families to perform surgical interventions [28, 29].

Medical models could be also advantageous for medical students and residents. These models offer a realistic demonstration of anatomical structures and complicated abnormalities, thus provide intuitive knowledge for surgeons and trainees. In addition, medical models could serve as educational tools for training surgical procedures. As a result, medical models can improve the knowledge and skills of medical students and practitioners [21, 24, 30].

Despite the efficiency of medical models in facilitating the diagnostic and treatment procedures, their use in routine clinical practice is hampered by some limitations. The first limitation is that creating medical models requires special costly materials and equipment. Therefore, it is not economically feasible to fabricate medical models for all cases. Furthermore, 3D printing of medical models requires time. Thus, these models are virtually unusable in emergency cases that necessitate urgent intervention. Another limitation is the maximum size of an object that a 3D printer can build. Regarding this issue, the whole body and extremely large organ models are not available yet. Finally, the restricted printing precision and imaging techniques' resolution limit the accuracy of medical models. In this case, the resolution of the imaging techniques is the main limiting factor, because the minimum image slice thickness of the computed tomography (CT) and magnetic resonance imaging (MRI) images is in the order of millimeters, which is much greater than the minimum layer thickness of the currently available 3D printers [8, 29]. Despite these limitations, the use of medical models in dental and maxillofacial treatments is increasing. The most widely used medical models in this field include dental models, TMJ models, mandibular models, facial models, and skull models.

4.2   Dental Models

3D anatomical dental models are robust assisting tools for performing complicated dental surgeries, such as tooth impaction, fused tooth, and tooth autotransplantation.

4.2.1   Application of Dental Models in the Management of Tooth Impaction

In 2006, Faber et al. [31] presented the case of a 13-year-old boy complaining of teeth misalignment. In the clinical examinations, class I malocclusion and maxillary left canine impaction were diagnosed. Since it is necessary to determine the exact position of an impacted tooth prior to the surgery, CT was performed. The CT

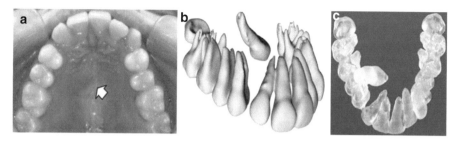

Fig. 4.1 (**a**) Intra-oral image of patient's maxillary teeth showing left canine impaction. The arrow points to the palatal mucosa protuberance caused by the impacted tooth. (**b**) CT image of the maxillary teeth. (**c**) The 3D printed model of the patient's maxillary teeth. Reproduced from [31] with permission

images (Fig. 4.1a) visualized the spatial relationship between the impacted tooth and the other teeth. For a more accurate diagnosis and preoperative planning, an anatomical model of the patient's maxillary teeth (Fig. 4.1b) was fabricated via photopolymer jetting (PPJ). The model made of UV-cured acrylic resin exhibited the exact anatomical relationships between the impacted canine and the adjacent structures. The 3D printed model was also used for intraoperative navigation during the surgery conducted to expose the tooth. Furthermore, the model reduced the surgery duration and facilitated communication with the patient and his family [31].

In 2013, Jang et al. [32] used 3D printing modeling to manage a complicated case of tooth impaction and autotransplantation. The 16-year-old female patient was diagnosed with a horizontally impacted immature molar tooth, with the adjacent tooth having been previously extracted to perform orthodontic treatments. The 3D anatomical dental model was fabricated and used for anatomical assessment and treatment planning. The treatment included extraction of the impacted tooth and its transplantation into the edentulous area. The 3D dental model facilitated the preoperative preparation of the recipient socket and allowed for immediate transplantation of the extracted tooth. Follow-up results confirmed the functionality of the transplanted tooth and its root development after transplantation [32].

4.2.2 Application of Dental Models in the Treatment of Fused Tooth

In 2009, Lucey et al. [33] reported the first use of 3D printed dental models in the management of fused teeth. An 11-year-old female patient was diagnosed with fused left maxillary lateral incisor and displaced adjacent canine. Cone-beam computed tomography (CBCT) imaging was performed to identify the exact location of the fusion. The images were employed to fabricate a real-sized 3D dental model via stereolithography (SLA) 3D printing. Based on the SLA model, the surgical planning was completed. The model was also used to educate and inform the patient and her family. Ten months after the surgery, the success of the treatment was confirmed in clinical examinations and radiographic assessments [33].

Kato and Kamio [34] presented a more complicated case of the fused tooth in 2015. A 16-year-old boy was referred for the treatment of a cold-sensitive tooth. Clinical examinations and intraoral radiographs revealed that the sensitivity was due to a tooth-like structure fused to the second mandibular molar. The treatment was extremely difficult because of the connected pulp cavities of the molar tooth and the supernumerary structure. Both endodontic and dental pulp treatment were necessitated. To better understand the morphology at the fusion site, CBCT was performed, however, the resulting images only represented the curvature of the root and its diameter variation. To overcome the limitations of CBCT, a 3D model was fabricated via fused deposition modeling (FDM). The 3D printed model provided a clear observation of the complicated structures and, thus, facilitated the diagnosis and treatment planning significantly by allowing for performing a simulated treatment and gaining informed consent from the patient [34].

4.2.3   Application of Dental Models in Dental Autotransplantation Surgeries

3D printed dental models have been used in dental autotransplantations. Some challenging requirements of these treatments include minimal damage to the periodontal ligament of the donor tooth, the maximum fit between the donor tooth and the recipient socket, and implantation of the donor tooth within a maximum of 30 minutes after the extraction. The key to success in autotransplantation is to minimize the extraoral time of the donor tooth. This is realized only if the socket is prepared prior to the extraction of the donor tooth [35, 36].

The first studies on the application of 3D printed dental models in autotransplantation surgeries were carried out by Lee et al. in 2001 [35]. Using anatomical resin dental models in pre-contouring of the recipient alveolar bone, they minimized the extraoral time of the donor tooth during the surgery. For 22 cases included in their study, the average total duration of the autotransplantation procedures was 7.7 min. The authors suggested using 3D printed dental models in autotransplantation of premolars to incisor sockets after traumatic tooth loss [35].

In 2010, Honda et al. [37] reported a dental autotransplantation case in which the right mandibular third molar of a 33-year-old male patient was grafted into the extraction socket of the left mandibular first molar. The bone volume of the socket and the dimensions of the graft tooth were determined by CBCT images (Fig. 4.2a). The stereolithographic real-sized model of the graft tooth was fabricated using urethane dimethacrylate (UDMA) photopolymerizable resin (Fig. 4.2b). Prior to the extraction of the graft tooth, the volume of the recipient socket was adjusted to be fitted with the size of the replica model. Furthermore, the contact of the replica tooth with the adjacent and opposite teeth was investigated. Using the replica model in cavity preparation reduced the time interval between the extraction and implantation of the graft tooth. Therefore, the graft tooth was implanted in the recipient socket immediately after being extracted, and as a result, the procedure was performed with minimal damage to the graft tooth and its periodontal tissues [37].

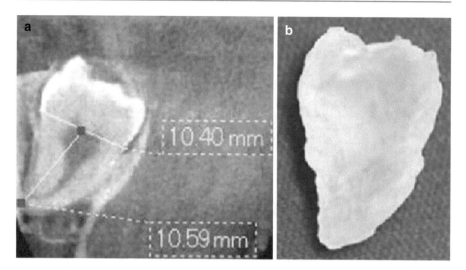

Fig. 4.2 (a) CBCT image of the graft tooth (right mandibular third molar). (b) Replica model of the graft tooth fabricated via SLA 3D printing. Reproduced from [37] with permission

In 2013, Jang et al. [32] performed four autotransplantation surgeries using third molars as the graft teeth. The cases aged between 15 and 21 years were diagnosed with prematurely lost molars. Autotransplantation was selected as the solution to replace the missing teeth. Using 3D printed dental models, the surgeries were carried out with a minimal extraoral time of the donor teeth. As the result, the periodontal ligament and Hertwig epithelial root sheath of the graft teeth remained vital. Periodic follow-up investigations of the transplanted teeth revealed normal pulps and developing roots. Besides, radiographic examinations of the transplanted teeth performed 2 years after the surgeries demonstrated increased periapical radiopacity. The success of the surgeries was confirmed, as no signs of transplant failures were observed after 7.5 years. The authors proposed 3D printing-assisted third molar autotransplantation as an alternative treatment for conventional implant-based approaches in patients with premature tooth loss [32].

In another study conducted by Verweij et al. in 2016 [38], autotransplantation was performed in young patients diagnosed with mandibular premolar agenesis. The orthodontic extracted maxillary premolars were transplanted into the neo-alveolus created at the recipient sites. In each case, a real-sized titanium replica of the donor tooth was created based on the 3D virtual model made from the CBCT images (Fig. 4.3). The replica was used for the preoperative preparation of the artificial socket. The donor tooth was extracted after the socket was prepared. Using 3D printed dental models as the surgical guides to obtain the maximum fit between the donor tooth and the neo-alveolus reduced the extra-alveolar time of the donor tooth to less than 1 min. As the result, the risk of the iatrogenic damage of the transplanted tooth was minimized. This study demonstrated that dental models fabricated by 3D printing could increase the success and survival rates of the auto-transplanted teeth [38].

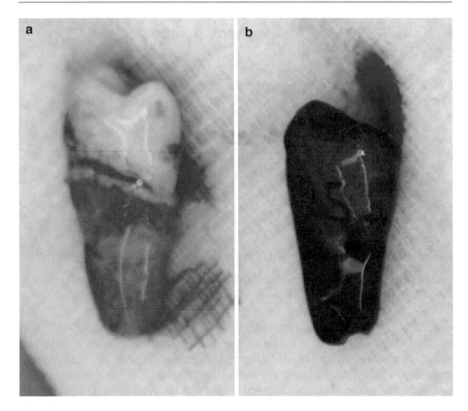

Fig. 4.3 (**a**) The orthodontically extracted premolar (the donor tooth). (**b**) The titanium replica of the donor tooth is used as the surgical guide. The figure is obtained from reference [37] with permission

4.3   TMJ Models

TMJ dysfunctions are caused by congenital deformities, traumatic injuries, tumors, and ankylosis. These implications impair mastication, swallowing, breathing, and speech. To restore the structure and function of TMJ, reconstruction surgeries are frequently needed, however, due to the complex structure and intricate anatomy of the TMJ, every surgical intervention in this region is highly challenging [39, 40].

3D printed TMJ anatomical models provide details of the hard and soft tissues of the TMJ in normal and pathological situations. These models make TMJ surgeries simpler and more convenient for the surgeon and the patient. In 2011, Mehra et al. [41] reported the reconstructive surgery of a 26-year-old male patient with facial asymmetry caused by an osteochondroma in his left TMJ (Fig. 4.4a). They fabricated the facial model of the patient via SLA and utilized it to determine the degree of the deformity (Fig. 4.4b). An accordingly appropriate treatment plan—including several complex osteotomy, chondyloctomy, and bone reconstruction procedures—was designed. Using the model, the surgeon successfully performed the simulated

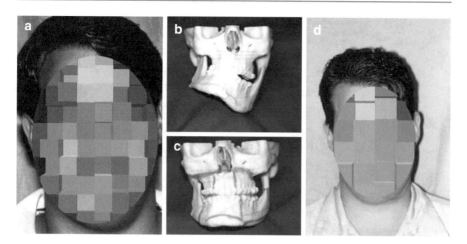

Fig. 4.4 (**a**) Patient with facial asymmetry caused by an osteochondroma in his left TMJ. (**b**) SLA facial model of the patient. (**c**) Facial model after the simulated surgery. Osteotomy segments are luted with wax in their new positions. (**d**) The same patient with significantly improved facial symmetry and contouring. Reproduced from [41] with permission

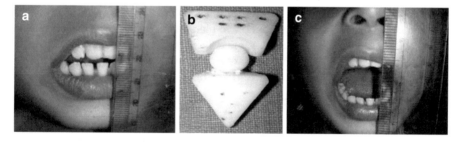

Fig. 4.5 (**a**) Limited mouth opening before the surgery. (**b**) ABS model of the TMJ implant fabricated via FDM. (**c**) Mouth opening immediately after the surgery. Reproduced from [42] with permission

surgery prior to the actual operation (Fig. 4.4c). The results showed that the symmetry and contouring of the patient's face improved significantly after the surgery (Fig. 4.4d) [41].

In another study, Deshmukh et al. [42] presented a case of TMJ ankylosis that led to reduced jaw opening (Fig. 4.5a). The patient's TMJ had broken due to the fall from height at the age of 6. The untreated ankylotic TMJ had impaired the patient's speech and mastication. Although no significant facial asymmetry or deformity was observed, the radiographic assessments revealed a bony adhesion on the left TMJ. Gap arthroplasty followed by customized prosthesis implantation was needed to restore the normal structure and functions of the joint. Using the FDM process, a customized model of the implant was fabricated from acrylonitrile butadiene styrene (ABS) (Fig. 4.5b) and used to plan the gap arthroplasty. Then, the surgical procedure was simulated by performing the required re-contouring of the fossa on

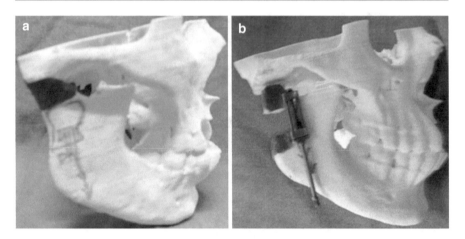

Fig. 4.6 (**a**) The SLA facial model presenting bony TMJ. (**b**) Preoperative planning to resect the ankylotic mass. Reproduced from [43] with permission

the ABS model. Finally, a metallic implant was fabricated from 316 L stainless steel using a vertical machining center (VMC). The implant fixation was performed on the ABS model to select the appropriate number and dimensions of the screws. The 3D printed model allowed for preoperative planning and simulated surgery, while also facilitating the design of the customized implant and its accurate and stable fixation. Immediately after the operation, the mouth opening was improved (Fig. 4.5c). Besides, normal jaw function was observed in postoperative follow-ups after 2 years from the surgery [42].

In 2018, Alwala et al. [43] performed a reconstruction surgery to manage bilateral TMJ ankylosis. The procedure included resection of the ankylotic mass and reconstruction of the TMJ via distraction osteogenesis. Preoperative planning was completed using a 3D anatomical model printed by SLA 3D printing (Fig. 4.6a) to be used for the identification of the ankylotic mass and planning the osteotomy procedure. The simulated surgery was also conducted on the model (Fig. 4.6b). The authors concluded that 3D printed anatomical models of the TMJ facilitate surgical interventions and improve the treatment outcomes of reconstructive operations [43].

4.4 Mandibular Models

One of the most commonly used medical models are mandibular models. These models are routinely used in corrective and reconstructive mandibular surgeries. In 2014, Prisman et al. [44] conducted a study to assess the efficacy of 3D printing-assisted contouring of mandibular plates. Ten patients with mandibular defects requiring reconstructive surgeries were included in the study. Real-sized SLA mandibular models were fabricated based on CT images and were used for preoperative contouring of mandibular plates (Fig. 4.7). Two experienced surgeons compared

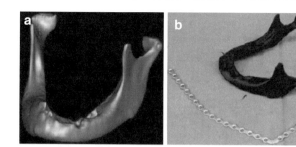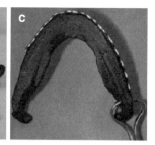

Fig. 4.7 (**a**) The virtual model is generated based on the CT images. (**b**) The SLA model and the titanium plate used in mandibular reconstruction. (**c**) The plate contoured preoperatively on the SLA model. Reproduced from [44] with permission

preoperatively- and intraoperatively-contoured plates in terms of an overall match of the plates, contour conformity, surface area of plate-mandible contact, and cost-effectiveness of the techniques. The surgeons were both blinded to the plating techniques. Overall, in seven surgeries, the SLA-assisted contoured plates were preferred to intraoperatively contoured plates. The authors concluded that preoperative mandible plate contouring is more advantageous compared to traditional intraoperative plating [44].

In 2017, Yoshimura et al. [45] used stereolithographic models in the reconstruction of condylar defects with vascularized fibular flaps. Five patients with condylar tumors were selected for mandibular resection. The defects were reconstructed using vascularized osteocutaneous fibular flaps. CT imaging was performed, and the data was used for creating mandibular models via SLA 3D printing. The SLA models were used for determining the lengths and angles of the bony reconstructions. The models also facilitated preoperative contouring of the fibular flaps (Fig. 4.8). To evaluate the morphology and functionality of the reconstructed mandibles, periodic follow-up examinations were performed for up to 23 months. The results showed significant improvement in mouth opening. The patients were able to return to their normal diet without any problem regarding occlusion. The aesthetic outcomes were also satisfactory, with no unfavorable outcomes observed, including bone resorption, ankylosis, or dislocation. The authors concluded that SLA models are useful tools for the reconstruction of condylar defects [45].

In 2018, Abbasi et al. [12] combined SLA 3D printing modeling with mirror imaging to reconstruct mandibular defects. After acquiring CT images, the image of the intact hemi-mandible was superimposed on the tumor-involved side. The resulting virtual model was used for fabricating a real-sized mandibular model via SLA 3D printing. The combined approach allowed for the preservation of the tumor structure on the SLA model. Therefore, the dimension of the tumor was exactly copied on the model. Using the SLA model, the resection area was determined, and osteotomy was performed with high accuracy and reliability. The stereolithographic model was then used for preoperative contouring of the reconstruction plate. According to the authors, the SLA models decrease the time needed for contouring

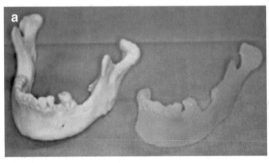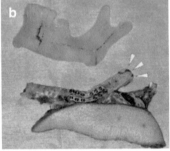

Fig. 4.8 (**a**) The SLA model of the mandible and a resin model created based on the SLA model. (**b**) The fibular flap was osteotomized according to the resin model. The arrows point to the end of the fibula, which has been roundly shaped to match the patient's glenoid fossa. Reproduced from [46], which is distributed under the terms of Creative Commons Attribution License

the reconstruction plates. Therefore, the problems associated with metal fatigue are eliminated, and the rates of implant failure are decreased [12].

4.5   Facial Models

The applications of 3D printed facial models in complicated maxillofacial surgeries are growing. Among these surgeries are facial asymmetry correction and orbital reconstruction. In 2009, Bell et al. [47] performed computer-assisted reconstruction surgeries in 15 patients with complex unilateral orbital deformities caused by trauma. Six patients had previously undergone conventional orbital restorative treatments. The imprecise restoration of orbital anatomy had led to unfavorable consequences, including facial asymmetry, diplopia, and/or enophthalmos. To improve the accuracy of the surgeries and avoid the negative postoperative sequellae, anatomical facial models were fabricated via SLA 3D printing. The traumatized side of each patient's face was reconstructed using the mirror image of the uninjured side. Surgical planning and preoperative contouring of the titanium meshes required to reconstruct the internal orbital wall were performed using the 3D printed models. Results showed that in the case of complex primary or secondary orbital deformities, preoperative anatomical modeling provides more accurate reconstruction with minimum unfavorable aftereffects [47].

In 2014, Novelli et al. [48] used SLA 3D printing modeling along with software planning and surgical navigation to perform orbitozygomatic reconstruction. Eleven patients with zygomatic deformity, unilateral orbital deformity, and associated diplopia and eno/exophthalmos were included in the study. Preoperative planning was carried out using real-sized stereolithographic models. Titanium meshes were used for orbital reconstruction and restoring healthy orbital volume (Fig. 4.9). Postoperative clinical examinations revealed complete correction of the eyeball position, and diplopia was resolved in all of the patients. The combinatorial approach used in the study had several preoperative and intraoperative advantages. First, it

Fig. 4.9 Preoperative mesh positioning and stabilization using the stereolithographic model. Obtained from reference [48] with permission

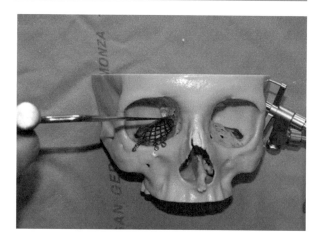

facilitated the diagnosis and treatment planning by providing a better understanding of the position and dimensions of the defects. Second, this made it possible to perform mesh positioning and stabilization preoperatively. Finally, this made the surgery simpler and expedited the intraoperative procedures [48].

In 2016, Kim et al. [49] reported surgical treatment of 16 patients with severe facial asymmetry caused by unilateral condylar overgrowth. The treatment included conservative condylectomy and simultaneous corrective orthognathic surgery. The amount of condylectomy was determined via radiographic assessments. To confirm the condylar overgrowth size and design a detailed surgical plan, stereolithographic facial models were fabricated based on the CT images. The models made it possible to predict the treatment outcome in advance. The postoperative results showed natural facial symmetry, and TMJ pain and limited mouth opening were also resolved after the surgery. Recurrence of condylar hyperplasia was not observed in long-term follow-ups. The authors suggested using stereolithographic models in complicated facial surgeries for in-depth investigation of patient anatomy and improved intraoperative accuracy [49].

In another study carried out in 2017, 82 patients with unilateral orbital wall fractures were selected to assess the accuracy of orbital reconstruction using 3D printed anatomical models. The fractures were planned to be repaired using porous polyethylene-coated titanium meshes. In the study group consisting of 44 patients, mirror-imaged stereolithographic models were fabricated and used for preoperative anatomical molding of the meshes (Fig. 4.10). In the remaining 38 patients, known as the control group, pre-contouring of the implants was not performed. Three months after the surgeries, the accuracy of implant insertion was measured in both groups. Postoperative surgical outcomes—including the degree of orbital dislocation, ocular motility, and diplopia—were also assessed. Results indicated a significant reduction of bone defect dimensions in the study group. Furthermore, the significantly lower values of gap lengths and layout angles confirmed the superior accuracy of implant insertion in the study group. Besides, the study group presented a lower rate of unfavorable postsurgical sequellae in comparison with the control

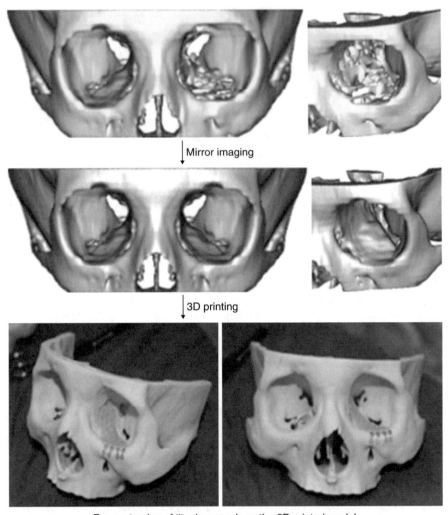

Pre-contouring of titanium mesh on the 3D printed model

Fig. 4.10 3D printing-assisted reconstruction of orbital fractures. Reproduced from [50] with permission

group. The authors concluded that patient-specific 3D printed anatomical models allow for more accurate restoration of orbital fractures [50].

4.6   Skull Models

Medical models have been successfully used in complicated skull surgeries. In 2012, Staffa et al. [51] exploited stereolithographic skull models in cranioplasty surgeries of 60 patients with large cranial defects. The defects were repaired using

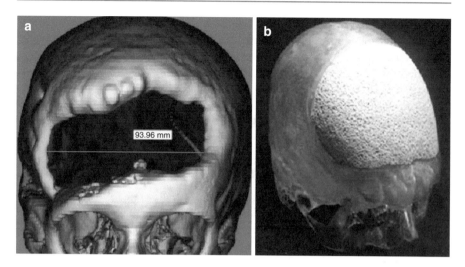

Fig. 4.11 (**a**) 3D CT image of a patient's skull representing the large frontal cranial defect. (**b**) The real-sized 3D printed skull model with the HA implant fitted at the defect site. Obtained from reference [51] with permission

custom-made porous hydroxyapatite (HA) implants. The process began by acquiring 3D CT scan images of patients' skulls. Then, the CT images were used to fabricate real-sized cranial models via SLA 3D printing (Fig. 4.11). The 1:1 acrylic models represented the anatomical position and dimensions of the defects and allowed for creating patient-specific HA prostheses accurately fitted to the defects. The SLA models facilitated the implantation process by removing the need for intraoperative manipulation of the implants. Two-year postoperative follow-up studies confirmed the safety and clinical performance of the implants. In addition, the aesthetic outcomes were favorable for all of the patients that participated in the study [51].

In another study, Martinez-Seijas et al. [52] proposed a novel method to fabricate PMMA cranial and orbital prostheses via stereolithographic models. Five patients with cranial and/or orbital bone defects, resulting from tumors or traumatic injuries, were included in the study. First, the position and dimensions of the defects were determined by CT imaging. The CT images were then used to print real-sized stereolithographic master models. The next step was to fabricate silicon injection molds based on the SLA master models, which were used to create patient-specific PMMA prostheses during the surgeries. The proposed method allowed for the construction of patient-specific PMMA cranial and orbital implants in a cost-effective manner. Follow-up studies conducted up to 7 years after the surgeries showed no complications, indicating the safety and compatibility of the injection-molded PMMA implants [52].

In 2019, Talamonti et al. [53] used SLA modeling in the reconstruction of a large cranial defect in an 18-month-old patient. CT imaging was performed and a sterilizable real-sized skull model was fabricated accordingly via SLA 3D printing

(Fig. 4.12). To reconstruct the defect, the SLA model was used intraoperatively to assemble and contour a customized three-piece bone flap (Fig. 4.13). An autologous bone layer obtained by craniotomy was used to cover the central part of the defect. The anterior and posterior portions of the defect were also covered with two homologous cadaveric bone layers. The flap was temporarily fixed using metallic plates and screws. CT images obtained 8 months after the surgery revealed complete engraftment of the implanted construct. The authors concluded that the proposed

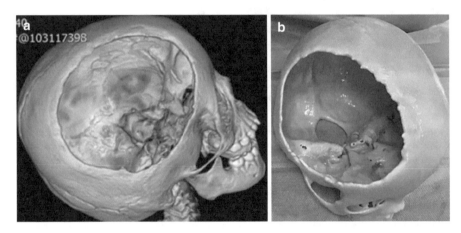

Fig. 4.12 (a) CT image of the 18-month-old patient with a large frontoparietotemporal cranial defect (defect size: 12 × 14 cm). (b) The real-sized stereolithographic model was fabricated based on the CT image. Reproduced from [53] with permission

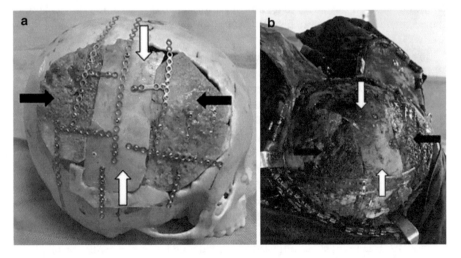

Fig. 4.13 (a) Assembling and contouring the bone flap on the SLA model. (b) Implantation of the bone flap. The white arrows point to the central piece of the flap (the autologous bone layer), and the black arrows show homologous cadaveric bone layers. Reproduced from [53] with permission

SLA 3D printing-assisted technique can be used to manage large cranial defects in very young patients [53].

4.7 Summary

Medical models fabricated via 3D printing have several advantages for patients and surgeons in the treatment of oral and maxillofacial defects. By providing unprecedented anatomical details, these real-sized models allow for the exact determination of the position and dimension of the defects and make it possible to access the defect and perform corrective interventions with minimum invasiveness for surrounding tissues. Among different 3D printing techniques, SLA is the one used most often for this purpose. It has been proven that using such models in the preoperative planning of complicated cases can improve the safety and efficacy of the treatments. These models are also useful for informing the patients about the status of their disease and the process of the treatment. Due to these benefits, the tendency among surgeons to use these models in surgical interventions has increased.

Accordingly, several studies have been conducted to assess the effectiveness of medical models in complicated dental and maxillofacial surgeries. It has been shown that dental models facilitate the diagnostic and therapeutic processes in challenging cases, such as tooth impaction, fused tooth, and dental autotransplantation. Similarly, studies have revealed that using 3D printed models for managing TMJ ankylosis, deformities, injuries, and tumors facilitates complicated surgical interventions. In addition, 3D anatomical models have been shown to improve the therapeutic and aesthetic outcomes of corrective and reconstructive mandibular surgeries. The efficacy of 3D printed facial models in correcting facial asymmetry and reconstructing orbital fractures has also been demonstrated. Furthermore, skull models have made it possible to precisely reconstruct large cranial defects even in very young patients. Nevertheless, the routine use of 3D printing medical models in daily practice is limited due to the high cost of the materials and equipment.

References

1. Bak M, Jacobson AS, Buchbinder D, Urken ML. Contemporary reconstruction of the mandible. Oral Oncol. 2010;46(2):71–6.
2. Mazzoli A. Selective laser sintering in biomedical engineering. Med Biol Eng Comput. 2013;51(3):245–56.
3. Lindner A, Rasse M, Wolf H, Millesi W, Eglmeier R, Friede I. Indications and use of stereolithographic skull reconstructions in oromaxillofacial surgery. Radiologe. 1995;35(9):578–82.
4. Lee S-J, Jang K-H, Spangberg LS, Kim E, Jung I-Y, Lee C-Y, Kum K-Y. Three-dimensional visualization of a mandibular first molar with three distal roots using computer-aided rapid prototyping. Oral Surg Oral Med Oral Pathol Oral Radiol Endod. 2006;101(5):668–74.
5. Nagori SA, Bhutia O, Roychoudhury A, Pandey RM. Immediate autotransplantation of third molars: an experience of 57 cases. Oral Surg Oral Med Oral Pathol Oral Radiol. 2014;118(4):400–7.

6. Nayar S, Bhuminathan S, Bhat WM. Rapid prototyping and stereolithography in dentistry. J Pharm Bioallied Sci. 2015;7(Suppl 1):S216.
7. Cervenka P, Witty C, Liacouras P, Crecelius C. Physical simulation models in oral and maxillofacial surgery: a new concept in 3-dimensional modeling for removal of impacted third molars. J Oral Maxillofac Surg. 2019;77:1125–9.
8. Winder J, Bibb R. Medical rapid prototyping technologies: state of the art and current limitations for application in oral and maxillofacial surgery. J Oral Maxillofac Surg. 2005;63(7):1006–15.
9. Feng Z, Dong Y, Zhao Y, Bai S, Zhou B, Bi Y, Wu G. Computer-assisted technique for the design and manufacture of realistic facial prostheses. Br J Oral Maxillofac Surg. 2010;48(2):105–9.
10. Sabol JV, Grant GT, Liacouras P, Rouse S. Digital image capture and rapid prototyping of the maxillofacial defect. J Prosthodont. 2011;138(4):310–4.
11. Zhao L, Patel PK, Cohen M. Application of virtual surgical planning with computer assisted design and manufacturing technology to cranio-maxillofacial surgery. Arch Plast Surg. 2012;39(4):309.
12. Abbasi A, Parvin M, Bashiri S. Innovative use of a stereolithographic model together with the mirror image technique to reconstruct a defect in mandibular continuity. Br J Oral Maxillofac Surg. 2018;56(9):887–9.
13. Müller A, Krishnan KG, Uhl E, Mast G. The application of rapid prototyping techniques in cranial reconstruction and preoperative planning in neurosurgery. J Craniofac Surg. 2003;14(6):899–914.
14. Wu C-T, Lee S-T, Chen J-F, Lin K-L, Yen S-H. Computer-aided design for three-dimensional titanium mesh used for repairing skull base bone defect in pediatric neurofibromatosis type 1. Pediatr Neurosurg. 2008;44(2):133–9.
15. Madrazo I, Zamorano C, Magallón E, Valenzuela T, Ibarra A, Salgado-Ceballos H, Grijalva I, Franco-Bourland RE, Guízar-Sahagún G. Stereolithography in spine pathology: a 2-case report. Surg Neurol. 2009;72(3):272–5.
16. Kobets AJ, Ammar A, Nakhla J, Scoco A, Nasser R, Goodrich JT, Abbott R. Virtual modeling, stereolithography, and intraoperative CT guidance for the optimization of sagittal synostosis reconstruction: a technical note. Childs Nerv Syst. 2018;34:965–70.
17. Chandramohan D, Marimuthu K. Rapid prototyping/rapid tooling–a over view and its applications in orthopaedics. Int J Adv Eng Technol. 2011;2:435–48.
18. Starosolski ZA, Kan JH, Rosenfeld SD, Krishnamurthy R, Annapragada A. Application of 3-D printing (rapid prototyping) for creating physical models of pediatric orthopedic disorders. Pediatr Radiol. 2014;44(2):216–21.
19. Jackson A, Ray LA, Dangi S, Ben-Zikri YK, Linte CA. 3D printing for orthopedic applications: from high resolution cone beam CT images to life size physical models. In: Medical imaging 2017: imaging informatics for healthcare, research, and applications. International Society for Optics and Photonics; 2017. p. 101380T.
20. Webb P. A review of rapid prototyping (RP) techniques in the medical and biomedical sector. J Med Eng Technol. 2000;24(4):149–53.
21. Hieu L, Zlatov N, Vander Sloten J, Bohez E, Khanh L, Binh P, Oris P, Toshev Y. Medical rapid prototyping applications and methods. Assem Autom. 2005;25(4):284–92.
22. Suzuki M, Ogawa Y, Kawano A, Hagiwara A, Yamaguchi H, Ono H. Rapid prototyping of temporal bone for surgical training and medical education. Acta Otolaryngol. 2004;124(4):400–2.
23. Marro A, Bandukwala T, Mak W. Three-dimensional printing and medical imaging: a review of the methods and applications. Curr Probl Diagn Radiol. 2016;45(1):2–9.
24. Jacobo OM, Giachero VE, Hartwig DK, Mantrana GA. Three-dimensional printing modeling: application in maxillofacial and hand fractures and resident training. Eur J Plast Surg. 2018;41(2):137–46.
25. Giannatsis J, Dedoussis V. Additive fabrication technologies applied to medicine and health care: a review. Int J Adv Manuf Technol. 2009;40(1–2):116–27.
26. Stoor P, Suomalainen A, Lindqvist C, Mesimäki K, Danielsson D, Westermark A, Kontio RK. Rapid prototyped patient specific implants for reconstruction of orbital wall defects. J Cranio-Maxillofac Surg. 2014;42(8):1644–9.

27. Malik HH, Darwood AR, Shaunak S, Kulatilake P, Abdulrahman A, Mulki O, Baskaradas A. Three-dimensional printing in surgery: a review of current surgical applications. J Surg Res. 2015;199(2):512–22.
28. D'Urso PS, Barker TM, Earwaker WJ, Bruce LJ, Atkinson RL, Lanigan MW, Arvier JF, Effeney DJ. Stereolithographic biomodelling in cranio-maxillofacial surgery: a prospective trial. J Cranio-Maxillofac Surg. 1999;27(1):30–7.
29. Rengier F, Mehndiratta A, Von Tengg-Kobligk H, Zechmann CM, Unterhinninghofen R, Kauczor H-U, Giesel FL. 3D printing based on imaging data: review of medical applications. Int J Comput Assist Radiol Surg. 2010;5(4):335–41.
30. Chan D, Frazier KB, Tse LA, Rosen DW. Application of rapid prototyping to operative dentistry curriculum. J Dent Educ. 2004;68(1):64–70.
31. Faber J, Berto PM, Quaresma M. Rapid prototyping as a tool for diagnosis and treatment planning for maxillary canine impaction. Am J Orthod Dentofac Orthop. 2006;129(4):583–9.
32. Jang J-H, Lee S-J, Kim E. Autotransplantation of immature third molars using a computer-aided rapid prototyping model: a report of 4 cases. J Endod. 2013;39(11):1461–6.
33. Lucey S, Heath N, Welbury R, Wright G. Cone-beam CT imaging in the management of a double tooth. Eur Arch Paediatr Dent. 2009;10(1):49–53.
34. Kato H, Kamio T. Diagnosis and endodontic management of fused mandibular second molar and paramolar with concrescent supernumerary tooth using cone-beam CT and 3-D printing technology: a case report. Bull Tokyo Dent Coll. 2015;56(3):177–84.
35. Lee SJ, Jung IY, Lee CY, Choi SY, Kum KY. Clinical application of computer-aided rapid prototyping for tooth transplantation. Dent Traumatol. 2001;17(3):114–9.
36. Kim E, Jung J-Y, Cha I-H, Kum K-Y, Lee S-J. Evaluation of the prognosis and causes of failure in 182 cases of autogenous tooth transplantation. Oral Surg Oral Med Oral Pathol Oral Radiol Endod. 2005;100(1):112–9.
37. Honda M, Uehara H, Uehara T, Honda K, Kawashima S, Yonehara Y. Use of a replica graft tooth for evaluation before autotransplantation of a tooth. A CAD/CAM model produced using dental-cone-beam computed tomography. Int J Oral Maxillofac Surg. 2010;39(10):1016–9.
38. Verweij JP, Moin DA, Mensink G, Nijkamp P, Wismeijer D, van Merkesteyn JR. Autotransplantation of premolars with a 3-dimensional printed titanium replica of the donor tooth functioning as a surgical guide: proof of concept. J Oral Maxillofac Surg. 2016;74(6):1114–9.
39. Felstead AM, Revington PJ. Surgical management of temporomandibular joint ankylosis in ankylosing spondylitis. Int J Rheumatol. 2011;2011:854167.
40. Okeson JP, Perez C, Fricton JR. Temporomandibular joint disorders, orofacial disorders. Springer; 2017. p. 145–57.
41. Mehra P, Miner J, D'Innocenzo R, Nadershah M. Use of 3-D stereolithographic models in oral and maxillofacial surgery. J Maxillofac Oral Surg. 2011;10(1):6–13.
42. Deshmukh T, Kuthe A, Chaware S, Bagaria V, Ingole D. A novel rapid prototyping and finite element method-based development of the patient-specific temporomandibular joint implant. Comput Methods Biomech Biomed Engin. 2012;15(4):363–70.
43. Alwala AM, Kasireddy SK, Nalamolu B, Malyala SK. Transport distraction osteogenesis in reconstruction of condyle: use of a 3D model for vector planning. J Maxillofac Oral Surg. 2018;17(3):276–80.
44. Prisman E, Haerle SK, Irish JC, Daly M, Miles B, Chan H. Value of preoperative mandibular plating in reconstruction of the mandible. Head Neck. 2014;36(6):828–33.
45. Yoshimura H, Matsuda S, Ohba S, Minegishi Y, Nakai K, Fujieda S, Sano K. Stereolithographic model-assisted reconstruction of the mandibular condyle with a vascularized fibular flap following hemimandibulectomy: evaluation of morphological and functional outcomes. Oncol Lett. 2017;14(5):5471–83.
46. Yoshimura H, Matsuda S, Ohba S, Minegishi Y, Nakai K, Fujieda S, Sano K. Stereolithographic model-assisted reconstruction of the mandibular condyle with a vascularized fibular flap following hemimandibulectomy: evaluation of morphological and functional outcomes. Oncol Lett. 2017;14(5):5471–83.

47. Bell RB, Markiewicz MR. Computer-assisted planning, stereolithographic modeling, and intraoperative navigation for complex orbital reconstruction: a descriptive study in a preliminary cohort. J Oral Maxillofac Surg. 2009;67(12):2559–70.
48. Novelli G, Tonellini G, Mazzoleni F, Bozzetti A, Sozzi D. Virtual surgery simulation in orbital wall reconstruction: integration of surgical navigation and stereolithographic models. J Cranio-Maxillofac Surg. 2014;42(8):2025–34.
49. Kim H-S, Kim J-Y, Huh J-K, Park K-H. A surgical strategy for severe facial asymmetry due to unilateral condylar overgrowth. Int J Oral Maxillofac Surg. 2016;45(5):593–600.
50. Kim YC, Jeong WS, Park T-k, Choi JW, Koh KS, Oh TS. The accuracy of patient specific implant prebented with 3D-printed rapid prototype model for orbital wall reconstruction. J Cranio-Maxillofac Surg. 2017;45(6):928–36.
51. Staffa G, Barbanera A, Faiola A, Fricia M, Limoni P, Mottaran R, Zanotti B, Stefini R. Custom made bioceramic implants in complex and large cranial reconstruction: a two-year follow-up. J Cranio-Maxillofac Surg. 2012;40(3):e65–70.
52. Martinez-Seijas P, Díaz-Galvis LA, Hernando J, Leizaola-Cardesa IO, Aguilar-Salvatierra A, Gómez-Moreno G. Polymethyl methacrylate custom-made prosthesis: a novel three-dimension printing-aided fabrication technique for cranial and/or orbital reconstruction. J Craniofac Surg. 2018;29(5):e438–40.
53. Talamonti G, Crisà F, Canzi G. Transplant of adult bone for reconstruction of a large post-traumatic cranial defect in a very young baby. Pediatr Neurosurg. 2019;54:218–22.

Application of 3D Printing in Production of Dental Instruments

5.1 Introduction

3D printing has made it possible to fabricate personalized surgical instruments and patient-specific implants. Such 3D printed and patient-specific instruments are very beneficial to be employed in deformity and reconstructive surgeries, dental surgeries, oral and maxillofacial surgeries, and orthognathic surgeries.

The usage of 3D printing in oral and maxillofacial surgery dates back to three decades ago. Since then, a variety of instruments and implants—including surgical guides, splints, crown and dentures, dental implants, oral and maxillofacial implants, and total jaw prostheses—have been successfully fabricated via 3D printing. Recently, low-cost 3D printers have been available, which have an influential role in the popularity of 3D printed instruments and implants. In-office fabrication of patient-specific devices and crowns is now a realistic option. Moreover, 3D printing has made it possible to create surgical instruments, which are easily modifiable according to a surgeon's preferences. Besides, it allows for the on-demand fabrication of sterile surgical instruments in remote locations [1, 2].

This chapter will describe the varieties of surgical instruments that can be made via 3D printing approach. For example, surgical guides made via 3D printing techniques can be used for dental implant surgeries, mandibular reconstruction, and tooth autotransplantation. Moreover, the chapter will elaborate on the applications of 3D printing method to fabricate splints, oral and maxillofacial implants, dental implants, crowns, and dentures. Possibilities of manufacturing total jaw prostheses along with the relevant examples will also be discussed at the end of the chapter.

© Springer Nature Switzerland AG 2021
L. Tayebi et al., *3D Printing in Oral & Maxillofacial Surgery*,
https://doi.org/10.1007/978-3-030-77787-6_5

5.2   3D Printing of Surgical Instruments for Dental and Maxillofacial Surgery

3D printing provides a variety of useful instruments for dental and maxillofacial surgeries. The most frequent applications of 3D printed instruments include dental implant surgeries, orthognathic surgeries, and midface and mandibular reconstruction [2].

5.2.1   3D Printing of Surgical Guides

Surgical guides have become one of the most useful tools in dental and maxillofacial surgeries. The major indications of 3D printed surgical guides are dental implant, orthognathic surgeries, and tooth autotransplantation [2, 3].

5.2.1.1 Application of 3D Printing Guides in Dental Implant Surgeries

In dental implant surgeries, drilling guides facilitate the orientation of the drillings, while positioning guides allow for the exact placement of the implants in their sites as determined in preoperative planning.

One of the growing applications of 3D printed surgical guides is in flapless implant surgeries. Compared to conventional open flap surgery, the flapless approach has several advantages, including decreased operation time, decreased patient discomfort, accelerated healing, and preservation of soft and hard tissues. However, since the surgeon is unable to visualize anatomic and vital structures, the risk of implant malpositioning and the risk of harming the tissues increases in this approach [4, 5]. Presurgical planning together with surgical guides minimize the disadvantages of flapless surgeries.

Several studies have demonstrated that using surgical guides improves the safety and accuracy of flapless implant placement. In 2008, Ersoy et al. [6] conducted a study to evaluate the accuracy of 94 implants placed using stereolithographic surgical guides, between 2005 and 2006. They measured the angular and linear deviations among the planned and placed positions of each implant. The results showed 1.22 ± 0.85 mm and 1.51 ± 1.0 mm linear deviations at the implants' neck and apex, respectively. The mean angular deviation was $4.9° \pm 2.36°$. No significant differences were observed between the flapless ($n = 41$) and flap ($n = 53$) groups. The authors concluded that SLA surgical guides could serve as precise tools for transferring the optimal implant position from the presurgical planning to the actual surgical side. Furthermore, these guides facilitate flapless implant surgeries.

In 2009, Ozan et al. [7] investigated the clinical accuracy of 110 implants placed by the use of different types of stereolithographic surgical guides. The data required fabrication of the guides derived from 3-dimensional CT scans. For each type of surgical guide—including tooth-supported, bone-supported, and mucosa-supported—the mean linear and angular deviations of the placed implants from the planned positions were measured. In tooth-supported guides, linear deviations at the apical region of the implants were significantly lower as compared to the other types. However, linear deviations at the neck of the implants were not significantly different among the study groups. Similarly, angular deviations obtained for the tooth-supported guides were the least. The authors concluded that tooth-supported surgical guides fabricated via stereolithography are more accurate than mucosa- or bone-supported types.

In 2013, Flügge et al. [8] proposed a workflow to simplify the fabrication procedure of implant surgical guides (Fig. 5.1). Briefly, they acquired patients' anatomic data in one session using an intraoral scanner and cone-beam computed tomography (CBCT). After being combined, the data were directly transferred to an in-office stereolithographic printer to fabricate the drilling guides with no loss of accuracy. During the second session, implant surgeries were performed using the fabricated tooth-supported drilling guides with metal inserts. The results of this study indicate that the proposed workflow allows for precise positioning and placement of dental implants while reducing the time and the related costs.

In 2015, Van de Wiele et al. [9] conducted a study to investigate the accuracy of implant surgeries performed by inexperienced surgeons using 3D printed surgical guides. They fabricated mucosa-supported drilling guides via stereolithography and used them to place 75 implants in 16 patients with completely edentulous jaws. They compared the results with the data obtained from 52 implant surgeries performed by an experienced surgeon in 12 patients. Apical, corneal, and depth deviations were measured in the study group and compared to the reference data. The results did not show any significant differences between the groups, suggesting that using stereolithographic surgical guides in completely edentulous jaws eliminates the influence of the surgeon's experience on the outcomes of the surgeries.

Although guided surgery is known as the most precise method for implant placement, few surgeons use it. This is mainly due to the high costs of the equipment. In a study, Whitely et al. [10] proposed a practical workflow to fabricate highly accurate and cost-effective surgical guides using affordable desktop 3D printers. They fabricated a half-arch guide via stereolithography and used it in a flapless surgery to replace the missing maxillary first molar in a 40-year-old patient. The authors concluded that the proposed method could help to increase the use of surgical guides

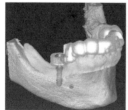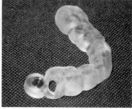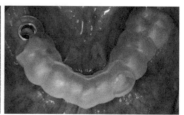

Fig. 5.1 Virtual drilling guide obtained by combining the data acquired from intraoral scanner and CBCT (Left). The printed drilling guide with metal insert and fenestrations to verify the implant placement process (Middle). The drilling guide is placed in situ (Right). Reproduced from [8] with permission

because it decreases the costs and eliminates the expenses of laboratory and shipping.

5.2.1.2 Application of 3D Printing Guides in Mandibular Reconstruction

In mandibular reconstructive surgeries, surgical guides facilitate the osteotomy procedures and prevent damage to dental anatomic structures. Cutting guides precisely determine the location and angulation of the osteotomy lines, while drilling guides identify the location for inserting the screws, and positioning guides determine the position of bone segments [2].

3D printing has been implemented in several studies to fabricate surgical guides needed for mandibular reconstruction. In 2014, Schouman et al. [11] implemented a complete digital workflow to perform mandibular reconstruction using free fibula flap (Fig. 5.2). First, the anatomical images were acquired and used to design the virtual planning and creation of 3D virtual maxillofacial and fibular models. Then,

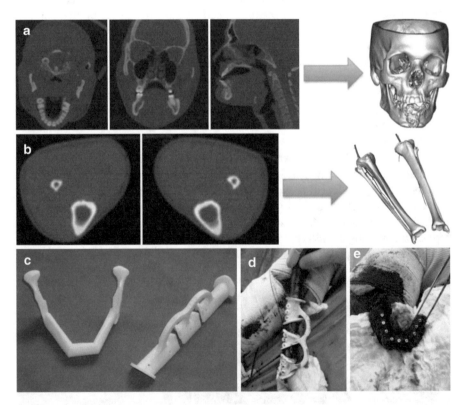

Fig. 5.2 3D virtual models created from segmented images: (**a**) Cranio-maxillary and mandibular model. (**b**) vascularized fibula model. (**c**) 3D printed cutting guide and fibular osteotomy template. (**d**) Fibular osteotomy using 3D printed template. (**e**) Placement of dental implants after fibular osteotomy. Reproduced from [11] with permission (Copyright ©2014, Elsevier Masson SAS. All rights reserved)

the models were transferred to a stereolithography 3D printer to fabricate customized cutting guides and fibular osteotomy templates. The customized 3D printed devices facilitated the defect resection, graft harvesting and forming, and contouring of the reconstruction plate. In addition, the proposed guided surgery approach allowed for concomitant reconstruction and prosthetic planning. Furthermore, it provided treatment outcomes with the best accuracy and reproducibility.

In another study, Liu et al. [12] proposed a technical procedure to reconstruct mandibular defects using 3D printed anatomical models and surgical guides (Fig. 5.3). The procedure consisted of three modules, including image module, defect module, and graft module. The image module used CT scanning to acquire

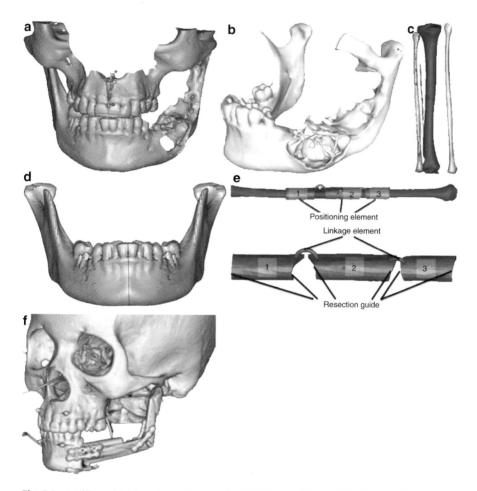

Fig. 5.3 (**a**) 3D model of maxilla and mandible. (**b**) The mandible model with the defect. (**c**) The model of shank bones and detached fibular bone. (**d**) Repaired model of the mandible after imposing image of the intact side on the defected side. (**e**) Designed resection template used for fibular osteotomy. (**f**) Postoperative image of the maxillofacial bones. Reproduced from [12] with permission from the author

the three-dimensional anatomical images of the mandible and fibula (Fig. 5.3a–c). The defect module allowed for the creation of the 3D model of the mandible with and without the defect (Fig. 5.3d). The defect and repair models were fabricated via stereolithography and used to perform defect resection and contouring of the reconstruction plate. Besides, a fibular resection template (Fig. 5.3e) was fabricated via the graft module, which made it possible to perform graft harvesting and shaping. The proposed modules provided customized models and tools for accurate transfer of the pre-operational virtual plan to the patient's defect site with more convenience. Furthermore, the operating time was decreased by 20% as compared with conventional surgeries.

Culié et al. [13] performed a retrospective study on 29 patients who underwent fibular free-flap mandibular reconstruction via conventional approach or virtual planning and guided surgeries (VPGS) in 2013 and 2014 (Fig. 5.4). The aim of the study was to compare the duration of surgeries and postoperative courses in the study groups. In the VPGS group, virtual planning was designed after acquiring maxillofacial and fibular CT images. Then, the customized cutting guides were fabricated via stereolithography. Although the duration of surgeries and postoperative course were not significantly different among the groups, the ischemia time of the fibula-free flap in the VPGS group was half of that in the conventional approach ($P < 0.001$). Another advantage of VPGS was the surgeon's comfort. Despite these advantages, the authors indicated that the high cost and the time required to obtain surgical guides limit the applicability of VPGS.

5.2.1.3 Application of 3D Printing Guides in Tooth Autotransplantation

Tooth replacement is challenging, especially in children, due to the growth and development of the alveolar bone. If an appropriate donor tooth exists, the preferred solution is dental autotransplantation. However, this kind of treatment is associated with challenges such as limited survivability of the donor tooth after extraction, the possibility of damage to the periodontal ligament of the donor tooth, and obtaining maximum fit between the donor tooth and the recipient socket [14]. 3D printing

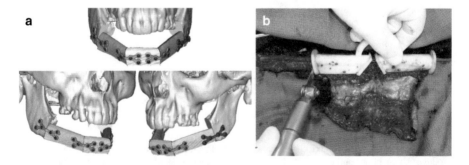

Fig. 5.4 (**a**) 3D modeling of mandibular reconstruction using free fibula flap. (**b**) Stereolithographic cutting guide positioned at the fibula. Reproduced from [13] with permission (Copyright ©2016, Elsevier Masson SAS. All rights reserved)

provides solutions to facilitate autotransplantation procedures by reducing the extraoral time of the donor tooth.

In 2010, Shahbazian et al. [15] conducted a study to investigate the accuracy and surgical feasibility of 3D printed surgical guides in tooth autotransplantation surgeries. They fabricated surgical guides with and without the replica tooth via stereolithography (Fig. 5.5). The surgical guides were used to prepare the recipient socket in a dry mandible model. To assess the accuracy of the replica, it was compared with the optical scan of the extracted tooth. For this purpose, the models were matched to each other, and point base analysis was performed to calculate the deviation between the surfaces. The results showed the accuracy level within 0.25 mm for 79% of the points. According to the authors, stereolithographic surgical guides provide satisfactory outcomes with acceptable accuracy. Therefore, this approach may improve the efficiency and predictability of autotransplantation surgeries.

In another study, the treatment outcomes of conventional autotransplantation surgeries were compared with stereolithography-assisted transfer techniques. For this purpose, 48 autotransplantation surgeries were carried out in 40 pediatric participants. In the study group ($n = 24$), CBCT images were acquired and used to fabricate donor teeth replicas and surgical guides (Fig. 5.6). The results demonstrated the advantages of the stereolithography-assisted approach over the conventional technique, including decreased extra-oral time of the donor tooth and reduced rate of postoperative complications. In addition, the guided approach protects the periodontal ligament of the donor tooth by eliminating the repetitive intraoperative

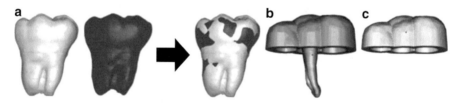

Fig. 5.5 (**a**) Comparison of the optical scans of the donor tooth (white) and the stereolithographic model (black). The design of the surgical guides with (**b**) and without (**c**) the replica tooth. Reproduced from [15] with permission

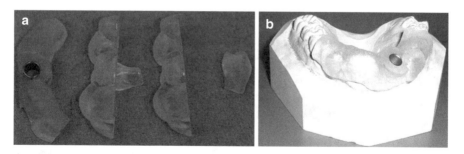

Fig. 5.6 (**a**) Stereolithographic surgical guides and tooth replica and (**b**) the study model. Reproduced from [16] with permission

manipulations needed to prepare the recipient socket in the conventional approach. The authors concluded that 3D printed surgical guides allow for less invasive auto-transplantation treatments and cause fewer failure rates [16].

5.3   Splints

Splints are among the most useful instruments used to treat tooth loosening, brux-ism, and occlusal disorders. In orthognathic surgeries, splints facilitate the reposi-tioning of the jaws in their correct anatomic locations to provide an optimized occlusion. They also allow for the transfer of the preoperative planning data to the surgical sites [17, 18].

The conventional manual methods of splint fabrication are laborious, time-consuming, and nonreproducible. In comparison, digital manufacturing methods—such as CAD/CAM and 3D printing—offer accuracy, speed, and consistency. Moreover, the digital techniques eliminate the need for impression, thus, provide more comfort to the patients during the process [19, 20].

In 2014, Adolphs et al. [21] assessed the clinical usability of occlusal splints fabricated via 3D printing. For this purpose, ten patients undergoing orthognathic corrective surgeries were included in the study. For each patient, a customized splint was fabricated through the PolyJet procedure (Fig. 5.7). Another splint was also made via the traditional method. Using RapidSplint® planning software platform, the optimal final occlusion for each patient was determined after acquiring the CBCT scans of the patient's dental casts. The surgeries were performed by immobi-lizing the dental arches; then, both splints were inserted, and the occlusal fitting was measured and compared. It was determined that eight of ten printed splints were usable to attain and preserve final occlusion. In five patients, the occlusal fitting of both types of splints was equal. In one case, the printed splint exhibited superior occlusal fitting as compared to the traditionally made splint. The study demon-strated that 3D printed splints are clinically usable to transfer preoperative planning data in orthognathic surgeries.

In 2017, Shaheen et al. [22] proposed a protocol to fabricate occlusal splints via 3D printing for 20 patients undergoing orthognathic surgeries. The printed splints

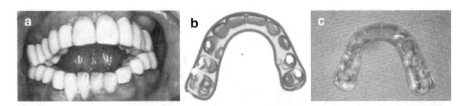

Fig. 5.7 (**a**) Preoperative situation of a patient with severe open bite. (**b**) The CAD model of the splint is specifically designed for the patient. (**c**) The 3D printed patient-specific splint. Reproduced from [21] which is distributed under the terms of the Creative Commons Attribution License (http://creativecommons.org/licenses/by/3.0)

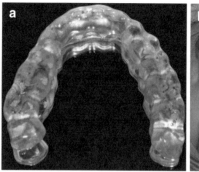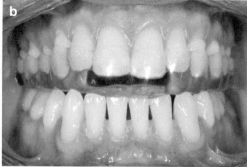

Fig. 5.8 (**a**) The 3D printed occlusal splint. (**b**) The splint inserted into the patient's mouth. Reproduced from [19] with permission

were analyzed in terms of dimensional accuracy. For this purpose, the preoperational planning was performed in PROPLAN software based on the CBCT data of the patients. Then, the splints were designed accordingly using 3-matic software. Afterward, the splints were made with a photopolymer jetting (PPJ) printer. To determine the accuracy of the printed splints, three landmarks were considered, and their distance errors from the same points on the conventionally made analog splints were measured/calculated and compared. The average distance error was 0.4 ± 0.17 mm, which is within the range of clinically acceptable error values. The study revealed that 3D printing allows for the fabrication of occlusal splints with clinically acceptable accuracy.

In another study, Waldecker et al. [19] recommended a complete digital procedure to manufacture patient-specific occlusal splints. The process begins by scanning mandibular and maxillary arches using an intraoral scanner. The centric relation of the arches is determined based on the acquired images. Then, the arrangement of the arches is virtually modified to obtain the optimal occlusion. Accordingly, an occlusal device is designed, and the model is transferred to a photopolymer printer (Fig. 5.8). After the fabrication, finishing, and polishing steps, the fit and handling of the final device are tested. The results of the study demonstrated that the proposed procedure allows for the production of functional occlusal splints with acceptable clinical accuracy and easy handling for patients.

5.4   Crowns and Dentures

Manufacturing dental restorations is a part of everyday dental practice. Ceramic restorations are conventionally fabricated through subtractive methods. The main disadvantage of the subtractive methods is material waste. Almost 90% of a zirconia block is wasted during the fabrication of a dental crown with a CAD/CAM milling machine [23]. Alternatively, this material waste is eliminated in 3D printing.

During the last decade, 3D printing has been successfully used for manufacturing dental restorations. A number of studies have focused on factors affecting the quality of the printed restorations. In 2016, Alharbi et al. [24] studied the effects of build angle and support thickness on the dimensional accuracy of dental prostheses fabricated via stereolithography. The dental crowns were printed in nine different angles, from 90° to 270°. For each build angle, printing was done using both a thin and thick support. To measure dimensional accuracy, the printed crowns were scanned using an optical surface scanner. The obtained models were compared with the scan data of the designed crown. The results showed the maximum accuracy by printing the crowns with thin support and a build angle of 120°. The improved accuracy of the crowns decreases the need for post-fabrication polishing and fishing. The study revealed that printing parameters affect the dimensional accuracy of dental restorations and proved that better results could be obtained by a smaller support surface area.

In 2017, Mai et al. [25] used photopolymer jetting (PPJ) to fabricate provisional dental crowns and compared the fit of the printed crowns with those manufactured via compression molding and milling. Different fabrication methods were used to manufacture 36 crowns ($N = 12$ for each group) (Fig. 5.9). Using silicone replica, the marginal, proximal, and internal (axial and occlusal) deviations between the restorations and the prepared teeth were measured with the image superimposition technique. The PPJ and milling crowns showed more accuracy in the proximal and marginal regions ($P < 0.001$). However, these crowns showed the largest deviations in the axial region. In the occlusal region, the accuracy of PPJ crowns was significantly higher than milling and molding crowns ($P < 0.001$). The study proved that PPJ technology provides a method to fabricate provisional crowns with high accuracy, especially in the occlusal region. Since the enhanced fit of the PPJ crowns improves their mechanical and biological performance, the authors suggested the use of PPJ as an alternative to conventional techniques for manufacturing interim crowns.

The number of studies on 3D printing of ceramic prostheses is limited. In 2019, Wang et al. [26] fabricated zirconia dental crowns with stereolithography and compared the accuracy of the printed crowns with crowns manufactured via CAD/

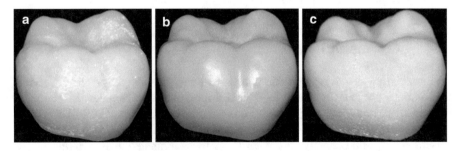

Fig. 5.9 Provisional crowns fabricated with molding (**a**), milling (**b**), and photopolymer jetting (PPJ). Reproduced from [25] with permission

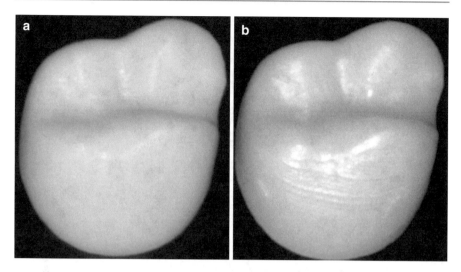

Fig. 5.10 Zirconia crowns fabricated with stereolithography (**a**) and CAD/CAM (**b**). Reproduced from [26] with permission

CAM. To fabricate the crowns, a maxillary second molar typodont tooth was used as the model (Fig. 5.10). The fabricated crowns ($N = 20$) were scanned, and the trueness analyses were performed by measuring the accuracy of external, marginal, intaglio, and intaglio occlusal surfaces. Results obtained for the studied surfaces revealed no significant difference between the accuracy of the 3D printed crowns and the CAD/CAM crowns ($P < 0.05$). The study showed that 3D printed zirconia crowns comply with the trueness requirements. Therefore, 3D printing could be considered an efficient method to fabricate zirconia crowns.

5.5   Patient-Specific Dental Implants

3D printing has been increasingly used to fabricate patient-specific dental implants. In the majority of the studies, implants have been printed using titanium and its alloys, however, zirconia [27, 28] and polyetheretherketone (PEEK) [29] have also been employed in some cases. The most commonly used methods to print the implants are the laser-based methods, including selective laser melting (SLM), direct metal laser sintering (DMLS), and direct metal laser forming (DMLF) [30–33].

3D printing has been implemented to manufacture both the endosteal and subperiosteal implants. Different types of endosteal implants—including threaded, non-threaded, solid, and porous—have also been fabricated via 3D printing [32, 34, 35].

The mechanical and biological performance of 3D printed dental implants has been evaluated in several studies. In 2014, Chen et al. [36] conducted a study to assess the dimensional accuracy and biomechanical performance of dental implants

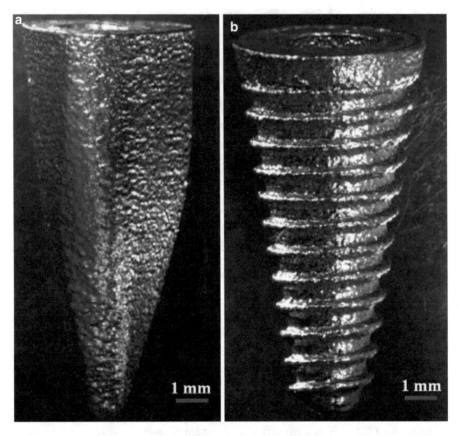

Fig. 5.11 Root-analog implant (left) and root-analog threaded implant (right), fabricated with selective laser melting. Reproduced from [36] with permission

fabricated via selective laser melting. First, computed tomography (CT) scans of a maxillary incisor were acquired. Then, the scans were used to create the virtual models of a root-analog and a root-analog threaded implant (Fig. 5.11). The models were transferred to an SLM printer and the implants were printed from grade two pure titanium powder. The dimensional accuracy of the implants was measured, while the mechanical performance of the implants was also evaluated in terms of tensile and bending strength. Besides, finite element analysis and resonance frequency analysis were carried out to assess the stress distribution and primary stability of the implants. The results revealed adequate dimensional accuracy and high strength for both types of implants. However, root-analog threaded implant showed better stress distribution, higher stability, and lower micromotions. The study proved that selective laser melting is an efficient method to fabricate patient-specific dental implants with adequate dimensional accuracy and high strength.

In 2016, a prospective 3-year follow-up study was performed by Tunchel et al. to assess the success and survival rates of 3D printed dental implants. Eighty-two patients indicated for single implant-supported restorations were included in the study. A total of 110 titanium implants were fabricated by direct laser forming, of which 65 implants were implanted in the maxilla and 35 implants in the mandible. After 3 years of loading, the radiographic, clinical, and prosthetic examinations were performed. The overall success and survival rates were 94.3% and 94.5%, respectively. Three years after loading, the average distance between the implant shoulder and the first bone-to-implant contact was 0.89 ± 0.45 mm. The study revealed that 3D printed dental implants are clinically successful for at least 3 years. The authors recommended performing further clinical investigations to assure the long-term performance and mechanical stability of the 3D printed dental implants.

3D printing allows for the fabrication of metallic and nonmetallic dental implants. In a study conducted in 2017 by Osman et al. [28], a custom-designed one-piece zirconia implant was printed via digital light processing (DLP) (Fig. 5.12). The dimensional accuracy, surface morphology, and mechanical properties of the printed implant were evaluated. The root mean square (RMSE) value of the implant was 0.1 ± 0.068 mm, indicating that the dimensional accuracy of the implant was adequate. In surface morphology examinations, cracks and interconnected nano- and micro-pores were found. Besides, the biaxial flexural strength test and Weibull analysis demonstrated that mechanical properties of the printed implant were in the range of the ceramic implants fabricated with conventional methods. The study revealed that DLP is capable of fabricating customized zirconia dental implants with adequate mechanical properties and dimensional accuracy. However, optimization of printing parameters is required to obtain implants with improved microstructure.

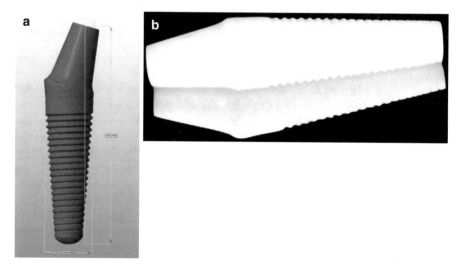

Fig. 5.12 (**a**) 3D model of the one-piece dental implant. (**b**) Zirconia implant printed with digital light processing. Reproduced from [28] with permission

Fig. 5.13 (a) The CAD model and (b) the 3D printed root-analog zirconia implant fabricated via DLP. Reproduced from [27] with permission

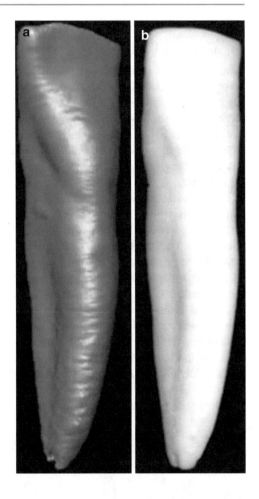

In a similar study, Anssari Moin et al. [27] assessed the feasibility of using DLP to fabricate zirconia root-analog implants. For this purpose, a cadaveric partially edentulous mandible was scanned, and the scan data was used to design the implant. The CAD model was then transferred to a DLP printer fed with commercial dental zirconia powder (Fig. 5.13). The accuracy of the 3D printed zirconia root-analog implant was evaluated by superimposing the CAD model, the surface model of the implant, and the extracted tooth scanned by an optical dental scanner. The deviations of the printed implant were within the clinically acceptable range. The maximum deviation (0.86 mm) was observed at the apical foramen. The authors concluded that the fabrication of one-piece zirconia root-analog implants with 3D printing is well feasible.

5.6   Oral and Maxillofacial Implants

In conventional approaches, occlusal wafers and splints are required for positioning and fixation of the maxilla after osteotomy. However, 3D printing provides a wafer-less solution and eliminates the need for splints. In 2015, Mazzoni et al. [37] developed a technique to fabricate patient-specific surgical guides and fixation plates to be used in maxillary repositioning. Ten patients with maxillofacial deformities were involved in the study to perform corrective surgeries. For each patient, CBCT scans were acquired and processed to create 3D virtual models of the maxillofacial hard and soft tissues. The models were used to design customized cutting guides and bone fixation plates (Fig. 5.14). The data were then transferred to a direct metal laser sintering machine to print the devices. Then, a comparison of virtually planned and post-surgically achieved positions of the maxilla was completed to assess the accuracy and reproducibility of the proposed method. The average accuracy was 92.7%, suggesting that the proposed waferless approach could be considered a promising method to fabricate customized surgical guides and fixation plates for maxillary corrective surgeries.

Another advantage of 3D printing is that the fabricated bone plates perfectly fit the maxillary and zygomatic contours of the patients. Due to the resulting "3D lock," the fitting of the plates and repositioned maxilla are possible only in one position. In 2016, Suojanen et al. [38] reported the use of 3D printing in manufacturing customized surgical instruments and implants for 32 patients undergoing maxillary corrective surgeries. After acquiring CBCT data and performing virtual planning, Le Fort I and bimaxillary osteotomies were conducted using 3D printed saw and drilling guides (Fig. 5.15). Then, titanium alloy fixation plates were fabricated individually for each patient. The plates were preoperatively bent based on the stereolithographic anatomical models of the patients. The results of the study specified precise positioning and fitting of the implants. The authors concluded that 3D printing allows for manufacturing patient-specific instruments and implants that are applicable in accurate and stable maxillary osteotomy and repositioning.

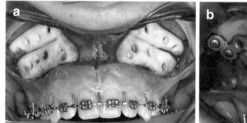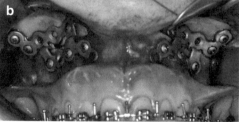

Fig. 5.14   (**a**) 3D printed patient-specific surgical guides and (**b**) fixation plates located at their determined positions. Reproduced from [37] with permission

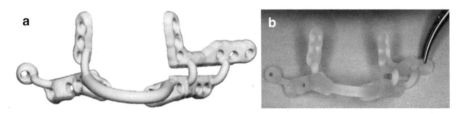

Fig. 5.15 Patient-specific saw and drilling guides: The CAD model (**a**) and the 3D printed version (**b**). Reproduced from [38] with permission

5.7   Total Jaw Replacement

Prosthetic joint replacement is the treatment of choice for severe TMJ degeneration when conservative therapies are not successful. It has been demonstrated that TMJ replacement decreases pain and discomfort while restoring tissue function in patients with tumors, trauma, osteoarthritis, and congenital abnormalities [39, 40]. Although different TMJ prostheses are commercially available, the shape and size of the components are not diverse enough to adapt to all patients' jaw morphologies [41].

3D printing allows for the fabrication of TMJ substitutes with customized size and geometry that conform to every patient's anatomy. In 2017, Ackland et al. [41] used 3D printing to fabricate a personalized prosthesis to replace the TMJ of a 58-year-old patient with severe osteoarthritis. The condylar component of the printed prosthesis was specifically sized to the patient. Besides, optimal screw positions were designed to facilitate prosthesis fixation and prevent possible damages to the mandibular nerve during the surgery. To evaluate the response of the prosthesis to mastication and maximum force bite, a customized musculoskeletal model was developed from the patient's medical images. A commercial TMJ prosthesis developed by Biomet Microfixation was also implanted and evaluated in terms of biomechanical performance (Fig. 5.16). Results obtained during mastication bite and maximum-force bite indicated lower mandibular stress, condylar stress, and screw stress for the printed prosthesis. Furthermore, 6 months after the surgery, the patient had normal mouth opening without any complications. The study suggested that 3D printing provides patient-specific TMJ prostheses with improved biomechanical and clinical performance than conventional stock devices. In addition, the risk of nerve damage during surgeries is decreased by the use of prostheses fabricated with 3D printing.

In a similar study [42], selective laser melting (SLM) was used to manufacture a patient-specific implant to substitute the degenerated TMJ of a 48-year-old female patient. The conservative TMJ therapies had previously failed for the patient. Besides, her joint anatomy was beyond the size and shape of the standard prostheses. The implant was designed in conformity to the patient's jaw anatomy acquired by CT scans (Fig. 5.17). The fossa of the implant was machined from high density polyethylene (HDPE), and the condylar component was printed from titanium

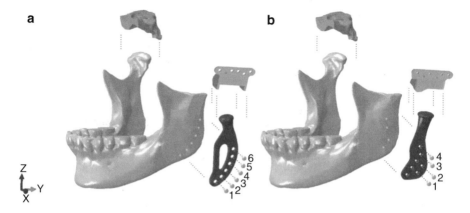

Fig. 5.16 (**a**) 3D printed TMJ prosthesis and (**b**) Biomet Microfixation prosthesis. Reproduced from [41] with permission

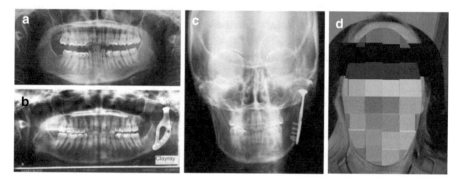

Fig. 5.17 Intraoral radiograph of the patient (**a**) before and (**b**) after the surgery. (**c**) Frontal view of patient's skull after implantation. (**d**) Increased mouth opening after the surgery. Reproduced from [42] with permission

6Al-4 V alloy. At 12 months after implantation, the functional performance of the implant was evaluated. Postsurgical clinical and radiographic examinations showed no complications. Besides, the results indicated an 86% reduction of joint pain and a 72% increase in mouth opening. The study demonstrated the effectiveness of the 3D printed TMJ prosthesis. The authors concluded that 3D printing has the prospective to fabricate patient-specific implants applicable in the treatment of patients with a wide range of end-stage TMJ degenerations.

In 2018, Dimitroulis et al. [43] conducted a prospective cohort study on 38 patients aged 20–66 years with end-stage joint degeneration that received TMJ prosthesis. The prostheses were fabricated using a combination of digital manufacturing platforms including CAD/CAM, computer numerical control (CNC), and 3D printing. With bilateral prostheses implanted in 12 of the patients, 50 implants were used in total. 80% of the prostheses were patient-matched, while the remaining 20% were completely customized (Fig. 5.18). The patient-matched implants were

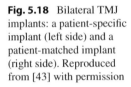

Fig. 5.18 Bilateral TMJ implants: a patient-specific implant (left side) and a patient-matched implant (right side). Reproduced from [43] with permission

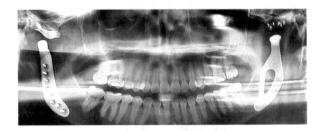

prepared in 1 week by sizing commercial implants to fit patients' anatomy. It took 3 weeks to prepare the completely customized implants, which were specially designed and manufactured to conform to each patient's unique anatomy. Results indicated that 12–24 months after total TMJ replacement, there were significant improvements in clinical indices. The joint function improved by 59.2%, while pain levels were reduced up to 74.4%. Improvement of patients' diet and mouth opening was 77.1% and 30.8%, respectively. During the study, no implant failures were observed. The study showed that a combination of additive and subtractive manufacturing methods could be implemented to rapidly deliver patient-matched and patient-specific TMJ implants. The proposed comprehensive digital platform provides safe and versatile substitutes to treat patients with end-stage TMJ degeneration.

5.8   Summary

Customized surgical instruments and patient-specific implants are becoming an integral part of maxillofacial surgeries and modern dental practices. 3D printing has played a major role in realizing this goal by allowing for in-office fabrication of surgical guides, splints, oral and maxillofacial implants, dental implants, crowns, and dentures. When these devices are fabricated via 3D printing, material waste is eliminated. Besides, the 3D printed instruments and implants are accurate and meet the surgeon's preferences. These advantages have been the reason for the increasing application of 3D printing in the fabrication of surgical instruments and implants. Although 3D printed devices are already commonly used in dental implant surgeries, orthognathic surgeries, and facial and mandibular reconstruction, it is expected that the scope of application of these devices will expand in near future.

References

1. Kondor S, Grant CG, Liacouras P, Schmid MJR, Michael Parsons L, Rastogi VK, et al. On demand additive manufacturing of a basic surgical kit. J Med Dev. 2013;7(3):030916.
2. Louvrier A, Marty P, Barrabé A, Euvrard E, Chatelain B, Weber E, et al. How useful is 3D printing in maxillofacial surgery? J Stomatol Oral Maxillofac Surg. 2017;118(4):206–12.
3. Tel A, Costa F, Sembronio S, Lazzarotto A, Robiony M. All-in-one surgical guide: a new method for cranial vault resection and reconstruction. J Cranio-Maxillofac Surg. 2018;46(6):967–73.
4. Oh TJ, Shotwell JL, Billy EJ, Wang HL. Effect of flapless implant surgery on soft tissue profile: a randomized controlled clinical trial. J Periodontol. 2006;77(5):874–82.

5. Oh T-J, Shotwell J, Billy E, Byun H-Y, Wang H-L. Flapless implant surgery in the esthetic region: advantages and precautions. Int J Periodont Restor Dent. 2007;27(1):27–33.
6. Ersoy AE, Turkyilmaz I, Ozan O, McGlumphy EA. Reliability of implant placement with stereolithographic surgical guides generated from computed tomography: clinical data from 94 implants. J Periodontol. 2008;79(8):1339–45.
7. Ozan O, Turkyilmaz I, Ersoy AE, McGlumphy EA, Rosenstiel SF. Clinical accuracy of 3 different types of computed tomography-derived stereolithographic surgical guides in implant placement. J Oral Maxillofac Surg. 2009;67(2):394–401.
8. Flügge TV, Nelson K, Schmelzeisen R, Metzger MC. Three-dimensional plotting and printing of an implant drilling guide: simplifying guided implant surgery. J Oral Maxillofac Surg. 2013;71(8):1340–6.
9. Van de Wiele G, Teughels W, Vercruyssen M, Coucke W, Temmerman A, Quirynen M. The accuracy of guided surgery via mucosa-supported stereolithographic surgical templates in the hands of surgeons with little experience. Clin Oral Implants Res. 2015;26(12):1489–94.
10. Whitley D III, Eidson RS, Rudek I, Bencharit S. In-office fabrication of dental implant surgical guides using desktop stereolithographic printing and implant treatment planning software: a clinical report. J Prosthet Dent. 2017;118(3):256–63.
11. Schouman T, Bertolus C, Chaine C, Ceccaldi J, Goudot P. Surgery guided by customized devices: reconstruction with a free fibula flap. Rev Stomatol Chir Maxillofac Chir Orale. 2014;115:28–36.
12. Y-f L, Xu L-w, Zhu H-y, Liu SS-Y. Technical procedures for template-guided surgery for mandibular reconstruction based on digital design and manufacturing. Biomed Eng Online. 2014;13(1):63.
13. Culie D, Dassonville O, Poissonnet G, Riss J-C, Fernandez J, Bozec A. Virtual planning and guided surgery in fibular free-flap mandibular reconstruction: a 29-case series. Eur Ann Otorhinolaryngol Head Neck Dis. 2016;133(3):175–8.
14. Kim E, Jung J-Y, Cha I-H, Kum K-Y, Lee S-J. Evaluation of the prognosis and causes of failure in 182 cases of autogenous tooth transplantation. Oral Surg Oral Med Oral Pathol Oral Radiol Endod. 2005;100(1):112–9.
15. Shahbazian M, Jacobs R, Wyatt J, Willems G, Pattijn V, Dhoore E, et al. Accuracy and surgical feasibility of a CBCT-based stereolithographic surgical guide aiding autotransplantation of teeth: in vitro validation. J Oral Rehabil. 2010;37(11):854–9.
16. Shahbazian M, Jacobs R, Wyatt J, Denys D, Lambrichts I, Vinckier F, et al. Validation of the cone beam computed tomography–based stereolithographic surgical guide aiding autotransplantation of teeth: clinical case–control study. Oral Surg Oral Med Oral Pathol Oral Radiol. 2013;115(5):667–75.
17. Warunek SP, Lauren M. Computer-based fabrication of occlusal splints for treatment of bruxism and TMD. J Clin Orthodont. 2008;42(4):227.
18. Metzger MC, Hohlweg-Majert B, Schwarz U, Teschner M, Hammer B, Schmelzeisen R. Manufacturing splints for orthognathic surgery using a three-dimensional printer. Oral Surg Oral Med Oral Pathol Oral Radiol Endod. 2008;105(2):e1–7.
19. Waldecker M, Leckel M, Rammelsberg P, Bömicke W. Fully digital fabrication of an occlusal device using an intraoral scanner and 3D printing: a dental technique. J Prosthet Dent. 2019;121(4):576–80.
20. Lauren M, McIntyre F. A new computer-assisted method for design and fabrication of occlusal splints. Am J Orthod Dentofac Orthop. 2008;133(4):S130–S5.
21. Adolphs N, Liu W, Keeve E, Hoffmeister B. RapidSplint: virtual splint generation for orthognathic surgery–results of a pilot series. Comput Aided Surg. 2014;19(1–3):20–8.
22. Shaheen E, Sun Y, Jacobs R, Politis C. Three-dimensional printed final occlusal splint for orthognathic surgery: design and validation. Int J Oral Maxillofac Surg. 2017;46(1):67–71.
23. Strub JR, Rekow ED, Witkowski S. Computer-aided design and fabrication of dental restorations: current systems and future possibilities. J Am Dent Assoc. 2006;137(9):1289–96.
24. Alharbi N, Osman RB, Wismeijer D. Factors influencing the dimensional accuracy of 3D-printed full-coverage dental restorations using Stereolithography technology. Int J Prosthodont. 2016;29(5):503–10.

25. Mai H-N, Lee K-B, Lee D-H. Fit of interim crowns fabricated using photopolymer-jetting 3D printing. J Prosthet Dent. 2017;118(2):208–15.
26. Wang W, Yu H, Liu Y, Jiang X, Gao B. Trueness analysis of zirconia crowns fabricated with 3-dimensional printing. J Prosthet Dent. 2019;121(2):285–91.
27. Anssari Moin D, Hassan B, Wismeijer D. A novel approach for custom three-dimensional printing of a zirconia root analogue implant by digital light processing. Clin Oral Implants Res. 2017;28(6):668–70.
28. Osman RB, van der Veen AJ, Huiberts D, Wismeijer D, Alharbi N. 3D-printing zirconia implants; a dream or a reality? An in-vitro study evaluating the dimensional accuracy, surface topography and mechanical properties of printed zirconia implant and discs. J Mech Behav Biomed Mater. 2017;75:521–8.
29. Mounir M, Atef M, Abou-Elfetouh A, Hakam M. Titanium and polyether ether ketone (PEEK) patient-specific sub-periosteal implants: two novel approaches for rehabilitation of the severely atrophic anterior maxillary ridge. Int J Oral Maxillofac Surg. 2018;47(5):658–64.
30. Lin W-S, Starr TL, Harris BT, Zandinejad A, Morton D. Additive manufacturing technology (direct metal laser sintering) as a novel approach to fabricate functionally graded titanium implants: preliminary investigation of fabrication parameters. Int J Oral Maxillofac Implants. 2013;28(6):1490–5.
31. Mangano F, Chambrone L, Van Noort R, Miller C, Hatton P, Mangano C. Direct metal laser sintering titanium dental implants: a review of the current literature. Int J Biomater. 2014;2014:461534.
32. Cerea M, Dolcini GA. Custom-made direct metal laser sintering titanium subperiosteal implants: a retrospective clinical study on 70 patients. Biomed Res Int. 2018;2018:5420391.
33. Figliuzzi M, Giudice A, Rengo C, Fortunato L. A direct metal laser sintering (DMLS) root analogue implant placed in the anterior maxilla. Case report. Ann Ital Chir. 2019;8:S2239253X19030044.
34. Mangano C, Bianchi A, Mangano FG, Dana J, Colombo M, Solop I, et al. Custom-made 3D printed subperiosteal titanium implants for the prosthetic restoration of the atrophic posterior mandible of elderly patients: a case series. 3D Print Med. 2020;6(1):1–14.
35. Oliveira TT, Reis AC. Fabrication of dental implants by the additive manufacturing method: a systematic review. J Prosthet Dent. 2019;122(3):270–4.
36. Chen J, Zhang Z, Chen X, Zhang C, Zhang G, Xu Z. Design and manufacture of customized dental implants by using reverse engineering and selective laser melting technology. J Prosthet Dent. 2014;112(5):1088–95.e1.
37. Mazzoni S, Bianchi A, Schiariti G, Badiali G, Marchetti C. Computer-aided design and computer-aided manufacturing cutting guides and customized titanium plates are useful in upper maxilla waferless repositioning. J Oral Maxillofac Surg. 2015;73(4):701–7.
38. Suojanen J, Leikola J, Stoor P. The use of patient-specific implants in orthognathic surgery: a series of 32 maxillary osteotomy patients. J Cranio-Maxillofac Surg. 2016;44(12):1913–6.
39. Giannakopoulos HE, Sinn DP, Quinn PD. Biomet microfixation temporomandibular joint replacement system: a 3-year follow-up study of patients treated during 1995 to 2005. J Oral Maxillofac Surg. 2012;70(4):787–94.
40. Sidebottom A, Gruber E. One-year prospective outcome analysis and complications following total replacement of the temporomandibular joint with the TMJ concepts system. Br J Oral Maxillofac Surg. 2013;51(7):620–4.
41. Ackland DC, Robinson D, Redhead M, Lee PVS, Moskaljuk A, Dimitroulis G. A personalized 3D-printed prosthetic joint replacement for the human temporomandibular joint: from implant design to implantation. J Mech Behav Biomed Mater. 2017;69:404–11.
42. Ackland D, Robinson D, Lee PVS, Dimitroulis G. Design and clinical outcome of a novel 3D-printed prosthetic joint replacement for the human temporomandibular joint. Clin Biomech. 2018;56:52–60.
43. Dimitroulis G, Austin S, Lee PVS, Ackland D. A new three-dimensional, print-on-demand temporomandibular prosthetic total joint replacement system: preliminary outcomes. J Cranio-Maxillofac Surg. 2018;46(8):1192–8.

3D Printing in Treatment of Soft, Hard, and Critical-Sized Oral and Maxillofacial Tissue Defects

6

6.1 Introduction

The oral and maxillofacial region consists of complex hard and soft tissues with individual anatomic discrepancies. Congenital deformities, tumors, and traumas can result in oral and maxillofacial defects that need interventions to be repaired. During the last two decades, significant advancements have been made in reconstructive surgeries; however, the management of oral and maxillofacial defects is still challenging due to the structural complexity and unlimited versatility of the defects. 3D printing provides a promising solution to address these challenges [1]. It has the potential to fabricate tissue constructs from different biomaterials and cells to resemble the architecture and structure of the defects. The printed constructs provide safe and effective tissue repair and minimize the role of the surgeon's skills on treatment outcomes [2]. However, this technology still needs more development to obtain optimized biomaterials and methods.

This chapter starts by discussing the tissue-engineering method as an effective approach to treat oral and maxillofacial defects. Three key factors in this method include stem cells, signaling molecules, and scaffolds, which will be discussed in further detail. 3D printing of oral and maxillofacial hard tissues will be described after analyzing the current treatment options to reconstruct hard tissue defects in the oral and maxillofacial region. Requirements and advances in 3D printing of oral and maxillofacial hard tissues will be discussed next. Similarly, for soft tissue reconstruction in the oral and maxillofacial region, 3D printing and 3D bioprinting as new approaches will be elaborated after reviewing the current treatment options.

The handling of critical-sized defects is a well-known major issue in oral and maxillofacial surgery. This chapter discusses different challenges in the treatment of such defects. Using the growing era of personalized medicine, by designing customized implants/scaffolds to fit each patient's needs, and also producing technologies that promote enhanced regeneration, such as growth factor delivery (especially for elderly patients) and electrical stimulation, we can make advances in the handling of critical-sized defects.

© Springer Nature Switzerland AG 2021
L. Tayebi et al., *3D Printing in Oral & Maxillofacial Surgery*,
https://doi.org/10.1007/978-3-030-77787-6_6

In this chapter, we will conclude that tissue engineering using customized 3D printed scaffolds loaded by growth factors can offer a good approach for the treatment of critical-sized defects. However, different strategies must be considered for various changes such as mechanical properties of scaffolds, vascularization, and healing time.

This chapter will introduce a design of a hybrid reinforced scaffold, composed of a 3D printed Skeleton filled with a Host Component (ref). The Skeleton is made of a nonporous material with strong mechanical properties and slow degradability, while the Host Component is made of a porous material that can be degraded faster. Thus, when the Host Component is degrading and being replaced by the regenerated tissue/cell, the Skeleton keeps the mechanical integrity of the cell/scaffold construct. The Skeleton can stay in place for a longer period of time and degrade very slowly over time until the treatment is completed (until the new bone gains enough mechanical strength after growing inside the porous Host Component). The design is inspired by reinforced concrete in which the metal rebars strengthen the structure of a concrete construct. In our design, the configuration of the 3D printed Skeleton can be optimized using finite element analysis based on the size, shape, and area (boundary condition) of the defect in the mandible. An example will be shown in this chapter, in which a bridge shape Skeleton could maximize the strength and minimize the material used. The last section of the chapter will deliberate the legal and ethical issues of tissue printing.

6.2   Tissue Engineering Triad: Stem Cells, Signaling Molecules, and Scaffolds

Tissue engineering is well known as a new approach for repair, replacement, or regeneration of tissues and organs that are damaged through traumatic injuries, aging, or illness. During the last two decades, tissue-engineering methods have had very promising outcomes in the regeneration of skin, cartilage, and, more specifically, bone. Three major factors for having a successful tissue-engineering procedure include stem cells, signaling molecules, and scaffold.

6.2.1   Stem Cells for Oral and Maxillofacial Tissue Regeneration

Selecting the appropriate stem cell is effective in oral and maxillofacial tissue regeneration. Different types of embryonic and adult stem cells have been suggested as potent cell sources applicable in oral and maxillofacial regeneration. Since using embryonic stem cells is ethically controversial, adult stem cells are preferred. Adult stem cells include mesenchymal stem cells (MSCs) and induced pluripotent stem cells (iPSCs). Clinical application of iPSCs is limited due to epigenetic memory and the risk of immune rejection and teratoma formation. Therefore, MSCs are often used in regenerative applications [3, 4].

MSCs are conventionally isolated and collected from bone marrow via aspiration technique. However, the process causes pain and morbidity at the donor site. Besides, the efficiency of the method in harvesting MSCs is low [5, 6]. To overcome the limits of bone marrow-derived mesenchymal stem cells (BMMSCs), several studies have been performed to harvest MSCs from other tissues. It has been determined that oral tissues—such as gingiva—could serve as efficient sources for MSCs' acquisition [3, 7].

The most common stem cell with intraoral origin applicable in regenerative purposes (Fig. 6.1) are as follows:

a. Gingiva-derived mesenchymal stem cells (GMSCs): GMSC is one of the easily accessible MSCs that can be simply isolated from the lamina propria of the gingiva. GMSCs exhibit long-term proliferative potential and retain their phenotype at high passages. They can differentiate into chondrocytes, osteoblasts, adipocytes, and neural cells [5, 9, 10].

b. Periodontal ligament stem cells (PDLSCs): PDLSCs are able to colonize on scaffolds and differentiate into chondrocytes, osteoblast-like cells, adipocytes, and neural cells. Besides, PDLSCs can express a great variety of cell surface markers. Although the growth of PDLSCs is faster than that of BMMSCs, BMMSCs possess greater bone-forming potential [11–13].

c. Dental follicle stem cells (DFSCs): Dental follicle is a mesenchymal connective tissue surrounding the developing teeth. The cementum, PDL, and alveolar

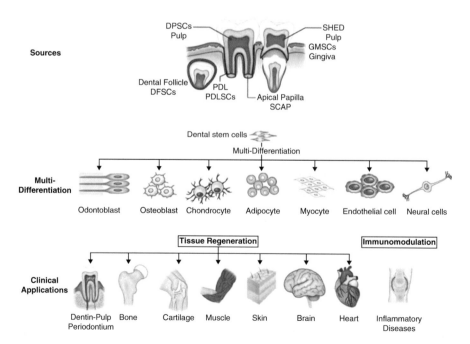

Fig. 6.1 Sources of stem cells with an intraoral origin, their differentiation capability, and potential clinical applications. Reproduced from [8] with permission

bone originate from the dental follicle. DFSCs are able to differentiate into cement-oblasts, osteoblasts, adipocytes, and neural cells [14, 15].

d. Dental pulp stem cells (DPSCs): DPSCs are multipotent stem cells that can be found at the perivascular areas of the pulp. DPSCs could be isolated from natal teeth, extra teeth, inflamed pulp, and extracted teeth. DPSCs are able to differentiate into chondrocytes, osteoblasts, odontoblasts, and neuronal cells [16, 17].

e. Stem cells from apical papilla (SCAP): SCAPs are able to differentiate into smooth muscle cells, chondrocytes, osteoblasts, odontoblasts, adipocytes, and neural cells. Premature roots of the third molars are the main sources to isolate SCAPs. It has been demonstrated that the proliferation rate of SCAPs is higher than that of DPSCs [18, 19].

f. Stem cells from human exfoliated deciduous teeth (SHED): SHEDs have the ability to differentiate into osteoblasts, odontoblasts, adipocytes, and neural cells. SHEDs are isolated from the pulp of the deciduous teeth and exhibit high differentiation potential and proliferation rate. Although SHEDs demonstrate osteoinductive property, they cannot promote bone formation directly due to the lack of osteogenic ability [20, 21].

6.2.2 Signaling Molecules Used for Oral and Maxillofacial Tissue Regeneration

Signaling molecules are another prerequisite for periodontal regeneration. The aim of using signaling molecules is to stimulate the stem cells to differentiate into desired lineages. The main signals used for oral and maxillofacial tissue regeneration are as follows:

a. Insulin-like growth factors (IGFs): IGF is an insulin-like molecule known to be potent in promoting bone formation. IGF has different isoforms. IGF-1 exhibits angiogenic and osteogenic properties while also able to promote cementum formation. Another isoform of IGF with known bone-forming property is IGF-2. IGF-2 triggers bone formation, but its efficacy is lower than IGF-1 [22]. To enhance periodontal regeneration, it has been suggested to use IGF-1 in combination with platelet-derived growth factor [23, 24].

b. Platelet-derived growth factor (PDGF): PDGFs regulate cell differentiation and proliferation. PDGFs exhibit mitogenic property on PDL cells and enhance the production of gingival fibroblast hyaluronate. The most commonly used member of the PDGF family, PDGF-B, is effective in repairing periodontal bone defects. It has been demonstrated that PDGF-B increases the attachment level and reduces gingival recession [25, 26].

c. Bone morphogenetic proteins (BMPs): BMPs are a well-known family of growth factors with angiogenic and osteogenic properties. Due to their morphogenetic potential, they stimulate the attachment and migration of stem cells onto scaffolds. It has been proven that bone morphogenetic protein-2 (BMP-2), bone morphogenetic protein-3 (BMP-3), and bone morphogenetic protein-7 (BMP-7) contribute to oral and maxillofacial tissue regeneration by inducing bone and carti-lage formation [27, 28].

d. Fibroblast growth factor (FGF): FGFs are a family of proteins with promi-nent differentiative and proliferative effects on mesenchymal cells. FGF signaling also plays a critical role in the formation of new blood vessels. FGFs are found in two forms, acidic FGF (FGF-1) and basic FGF (FGF-2). The most commonly used FGF for regenerative purposes is FGF-2. FGF-2 possesses angiogenic and mito-genic properties. It promotes periodontal regeneration by inducing new alveolar bone and cementum formation and facilitates soft tissue regeneration by enhancing the proliferation and migration of ligament cells [29–31].

e. Transforming growth factor-beta (TGF-β): TGF-β is a multifunctional growth factor that exhibits modulatory effects on proliferation, migration, and dif-ferentiation of various cell types. It triggers ECM production, bone matrix deposi-tion, and periodontal ligament fibroblast proliferation while also inducing chemotaxis of osteoblasts. TGF-β is a member of the TGF superfamily with three isoforms in mammalian species, including β1, β2, and β3. TGF-β1 enhances the proliferation of MSCs, preosteoblasts, osteoblasts, and chondrocytes. In addition, it promotes ECM formation and inhibits inflammation [32–35].

f. Enamel matrix derivative (EMD): EMD is a protein complex composed of amelogenin, amelotin, ameloblastin, enamelin, and tuftelin. Synthesized by amelo-blasts, EMD contributes to enamel mineralization by promoting the formation and growth of hydroxyapatite crystals. EMD also plays a key role in the development of root and periodontal attachment apparatus. Moreover, it induces differentiation of several MSCs including osteoblasts. EMD represents a promising biologic appli-cable in the reconstruction of complicated oral defects. During the last two decades, it has been increasingly used for the regeneration of oral and maxillofacial hard and soft tissues [36–38].

g. Periodontal ligament-derived growth factor (PDL-CTX): PDL-CTX is a novel growth factor isolated from periodontal cells. PDL-CTX is highly specific to PDL cells and has no chemotactic effects on gingival fibroblasts or epithelial cells. Therefore, its efficacy for periodontal regeneration is as much as 1000 times greater than that of IGF, PDGF, and TGF. For this reason, some studies suggest that any biological therapeutics used for oral and maxillofacial tissue regeneration should contain PDL-CTX [39–41].

h. Platelet-rich plasma (PRP): An easily accessible source of growth factors is platelet-rich plasma (PRP). PRP is a concentrated source of plasma proteins and autogenous growth factors, including epidermal growth factor (EGF), vascular endothelial growth factor (VEGF), IGF, and FGF-2. PRP is obtained by centrifug-ing the fresh blood of the patient. Several PRP kits are commercially available that allow for chair-side isolation and preparation of PRP. The modulatory effect of PRP on the proliferation of several periodontal cells has been demonstrated in vitro. It stimulates DNA synthesis in osteoblasts, periodontal ligament cells, and gingival fibroblasts. Moreover, it has also been reported that the proliferative effect of EMD is greater than that of single constituent growth factors [42, 43].

6.2.3   Scaffolds Used for Oral and Maxillofacial Tissue Regeneration

The third component of the tissue engineering triad is the scaffold. Scaffolds are required to provide an appropriate microenvironment for the proliferation and migration of cells. Scaffolds are prepared in different forms including hydrogels, sponges, and porous networks. Several natural and synthetic materials are used to fabricate scaffolds [44, 45].

The essential requirements of scaffolds include biocompatibility, biodegradability, porosity, and appropriate mechanical properties. The biocompatibility of the scaffold allows for attachment, migration, differentiation, and proliferation of the cells without any adverse effect on their viability. Scaffold is also necessary to eliminate severe inflammatory responses after the implantation of the scaffold. The biodegradability suggests that the degradation rate of the scaffold should proportionate with the rate of the new tissue formation. The porosity of the scaffold provides an interconnected porous structure to facilitate the penetration of the cells. It is also necessary to allow for the diffusion of the nutrients and growth factors within the construct and discharge of the cellular metabolites. Eventually, appropriate mechanical properties are required for the consistency of the scaffold with the surrounding tissues. Added to these requirements is the capability of the scaffold to support the formation of new vessels over its entire volume. A balance between porous structure, vascularization, and mechanical properties guarantees the success of any scaffold [46–48].

Varieties of biocompatible metals, ceramics, and polymers have been employed to fabricate scaffolds for the regeneration of oral and maxillofacial tissues [49]. The most commonly used scaffold materials are summarized as follows.

a. Metals and alloys: Since metals and alloys exhibit high strength, they are routinely employed as scaffolds for the regeneration of hard tissues. Titanium and its alloys—nitinol, stainless steel, cobalt–chromium alloys, tantalum, and magnesium alloys—are the most commonly used biocompatible metals and alloys. Magnesium (Mg), as a degradable metal, has also attracted attention for bone tissue engineering, however, due to the problems associated with the corrosion of this metal, Mg constructs have to be treated by effective coatings, or other methods to be applicable for bone regeneration [50–53]. The main criteria to be considered for the selection of a metal as a scaffold material are biocompatibility, appropriate mechanical properties, and low corrosion rate after implantation [54–56].

b. Ceramics: Ceramics have been used for two decades to fabricate dental and joint implants. Hydroxyapatite (HA) and beta-tricalcium phosphate (TCP) are the most widely used ceramics for bone tissue engineering. They exhibit biocompatibility and osteoinductivity while being able to not trigger immunogenic responses. HA is naturally found as the mineral component of bones, dentine, and enamel. It can be either synthesized or derived from allogenic or xenogenic sources. TCP is a bioresorbable ceramic with osteogenic potential. It has been reported that biphasic calcium phosphate (BCP)—which is the mixture of HA and TCP—outperforms its constituents in bone regeneration applications [57, 58]. Other ceramics and glasses such as bioactive glasses and zirconium titanate have also shown good results for bone tissue engineering [59–62].

   c. Polymers: Several natural and synthetic polymers are extensively employed as scaffold materials due to their biocompatibility, biodegradability, and processability. Natural polymers used in tissue engineering are ECM-derived proteins or polysaccharides. Collagen, fibrin, alginate, and chitosan have been used individually or in combination with other biomaterials to fabricate three-dimensional porous scaffolds. Although natural polymers are highly interesting for their inherent biocompatibility and biodegradability, their mechanical properties are not sufficient for load-bearing applications. Synthetic polymers are the alternatives to overcome this limitation. Polycaprolactone (PCL), polyethylene glycol (PEG), polyanhydride, polycarbonate, and polyesters are routinely used for scaffold fabrication [63–65].

6.3   3D Printing in Oral and Maxillofacial Hard Tissue Regeneration

6.3.1   Treatment Options to Reconstruct Oral and Maxillofacial Hard Tissue Defects

The gold standard for the reconstruction of oral and maxillofacial hard tissue defects is bone grafting. Bone grafting is extensively used in the mandibular reconstruction and corrective surgeries of craniofacial deformities. Another well-established application of bone grafting is for alveolar bone augmentation, which is performed to provide adequate bone support around the dental implants. Bone grafts are also used as fillers in post-extraction alveolar sockets, peri-implantitis-induced defects, and bone defects caused by trauma or infection [2].

   Bone grafts possess osteoinductive as well as osteoconductive properties and provide structural support for the overlying soft tissues. During the healing process, the graft is resorbed and replaced with new bone. Different types of bone grafts include:

- Autografts: Grafts from the own body of the recipient. Autologous bone grafts are commonly taken from a patient's iliac crest or fibula. Although autografts eliminate the risk of infection and immune rejection, they cause pain and morbidity at donor sites. Moreover, they are not easily available in children and elderly patients [66].
- Allografts: Grafts from other members of the same species. The most common forms of bone allografts are mineralized or demineralized freeze-dried grafts. Despite the rigorous chemical and radiative treatments, allografts may be prone to infection transmission [67].
- Xenografts: Grafts from a donor of a different species from the recipient. Bovine, porcine, and equine bone grafts are the most common forms of xenografts. Xenografts may facilitate the transmission of zoonotic diseases to the recipient [66, 67].

- Synthetic grafts: Synthetic grafts, although known as alloplastic grafts, do not possess osteogenic or osteoinductive properties. The main advantages of these grafts include structural integrity as well as the elimination of donor site morbidity and disease transmission [66, 67].

Despite its widespread application and significant outcomes, bone grafting is not applicable in the reconstruction of all types of hard tissue defects. For instance, reconstruction of delicate anatomical structures in the midface region is not possible by conventional bone grafting and requires particular strategies to mimic the complex shape and dimension of the bones [2].

6.3.2   Requirements and Advances in 3D Printing of Oral and Maxillofacial Hard Tissues

A promising alternative for bone grafting is tissue engineering via 3D printing. Progress in medical imaging and processing, digital manufacturing, smart biomaterials, and cell culturing techniques provided the basis for the emergence and advancement of 3D printing. In 3D printing, individualized tissue substitutes are precisely fabricated from materials of choice based on computer-aided designed anatomical models [68].

3D printing makes it possible to repair craniofacial, mandibular, and oral bone defects with anatomically precise features. It has the potential to resolve the challenges posed by diverse anatomical discrepancies of individuals. Furthermore, the use of printed constructs facilitates the surgical procedure and reduced the operation duration [69].

Until the last decade, printable materials were limited to metals and ceramics. Further developments permitted printing polymers and composites [70]. Recently, it has been possible to print a variety of stem cells combined with biomaterials and growth factors. In the fabrication of maxillofacial tissue constructs, oral stem cells, including DPSCs, SHEDs, PDLSCs, SCAPs, and DFPCs, are the main source of choice. It has been demonstrated that stem cells derived from oral tissues have the potential to regenerate functional hard and soft tissues [71, 72].

Maintaining the function of the stem cells is essential for further regenerative applications. Direct injection of the cellular suspension at the defect site has several restrictions for clinical application. First, mechanical forces exerted during injection destruct cell membranes and decrease cellular viability. Moreover, the absence of a 3D matrix results in low cell retention after transplantation [73, 74].

To overcome these limitations, appropriate 3D carriers are needed to provide both the mechanical and the biological requisites of stem cell-based regenerative approaches. For this purpose, several injectable and noninjectable hydrogels have been developed that allow for the safe and effective delivery of the cells. Cells are co-printed within the hydrogel or seeded after gelation. Hydrogels also permit the incorporation of growth factors and bioactive molecules [75].

The hydrogels are prepared from naturally derived or synthetic polymers [76]. The most commonly used natural polymers include collagen, hyaluronic acid, fibrin, chitosan, agarose, alginate, gelatin, silk, and cellulose. The favorable properties of natural polymers are similar to the ECM, hydrophilicity, biocompatibility, and biodegradability. The main drawback of these polymers is their weak mechanical properties. PCL, polylactide acid (PLA), polyglycolic acid (PGA), poly-lactic-co-glycolic acid (PLGA), polyethylene glycol (PEG), methacrylated gelatine, and Pluronic®-127 are the most frequently used synthetic polymers used for the 3D printing of tissue substitutes [77, 78].

In 2017, Ma et al. [79] used PDLSCs encapsulated in injectable hydrogels made of UV-crosslinkable gelatin methacrylate (GelMA) and poly(ethylene glycol) dimethacrylate (PEGDA) to repair alveolar bone defects. They implemented bioprinting for the encapsulation and delivery of cells. In addition, they studied stem cell–ECM interactions to find an appropriate ECM for the in vivo repair of the defects. Based on the results, 4:1 GelMA/PEGDA hydrogels exhibited maximum osteogenic differentiation in vitro and new bone formation in vivo. The study showed that bioprinting paves the way for the regeneration of functional alveolar bone by precise control of the ECM composition and optimizing cell–ECM interactions.

In 2018, You et al. [80] developed alginate/hydroxyapatite (ALG/HA) composite hydrogels to be implemented in the regeneration of osteochondral tissues. The ALG/HA constructs showed favorable cytocompatibility with chondrocytes. In vitro studies revealed that the viability and proliferation of chondrocytes increase by the addition of HA into the ALG hydrogels. Moreover, HA promoted the secretion of the mineralized matrix from chondrocytes. The ALG/HA hydrogel-upregulated cartilage markers include Collagen X and alkaline phosphatase. Besides, ALG/HAP constructs raised the mineralization level of chondrocytes after subcutaneous implantation in mice. To evaluate the printability of the ALG/HA hydrogels, porous structures were printed using a 250-µm diameter needle under 0.3 bar pressure (Fig. 6.2). Results showed uniform dispersion of HA particles within the ALG

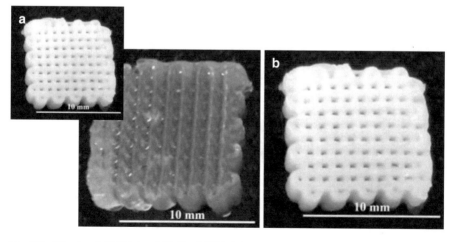

Fig. 6.2 3D printed ALG (**a**) and ALG + 1%HA (**b**) porous hydrogel scaffolds. Reproduced from [80] with permission

hydrogels. No clogging was observed during the printing of HA-added ALG hydrogels. The study demonstrated that the addition of HA into ALG hydrogels stimulates the secretion of a calcified matrix from chondrocytes. Since the HA/ALG hydrogels showed appropriate printability, they have the potential for bioprinting of osteochondral constructs.

In 2019, Chen et al. [81] developed a composite inorganic/biopolymer hydrogel with the desired mechanical and biological properties for bone regeneration. The hydrogels were prepared via photo crosslinking of methacrylated gelatin (Gel) and octamethacrylated polyhedral oligomeric silsesquioxane (POSS) nanocages. The Gel-POSS composite hydrogels had higher strength and a more favorable degradation rate and biocompatibility when compared to the conventional hydrogels. In vitro studies showed efficient adhesion, migration, and proliferation of rat bone marrow mesenchymal stem cells (rBMSCs) on Gel-POSS hydrogels. To evaluate the regenerative potential of the hydrogels in vivo, rBMSC-seeded structures were transplanted into the 5-mm rat calvarial defects. Micro-CT images obtained 8 weeks post-implantation revealed significant bone repair in the defects transplanted with Gel-POSS hydrogels (Fig. 6.3). The results of the study suggest that inorganic/biopolymer hydrogels with tailorable mechanical and biological properties provide promising platforms for stem cells.

In another study, Fahimipour et al. [82] developed biomimetic constructs through 3D printing to be used for personalized bone regeneration. The hybrid structure of the constructs was composed of cell-laden collagen hydrogels injected into 3D printed β-tricalcium phosphate (β-TCP) scaffolds (Fig. 6.4). The ECM-mimetic hydrogels were prepared from collagen conjugated with heparin. Prior to gelification, the collagen–heparin solution was mixed with recombinant human bone morphogenic protein 2 (rhBMP2). Then, dental pulp mesenchymal stem cells (MSCs)

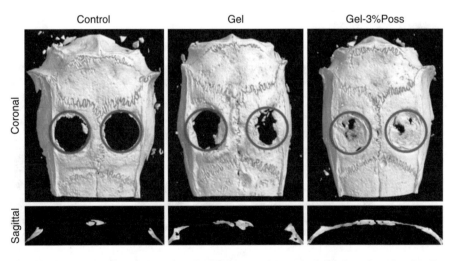

Fig. 6.3 Coronal (up) and sagittal (down) μCT images of the calvarial defects 8 weeks after the implantation of Gel and Gel-POSS hydrogels. Reproduced from [81] with permission

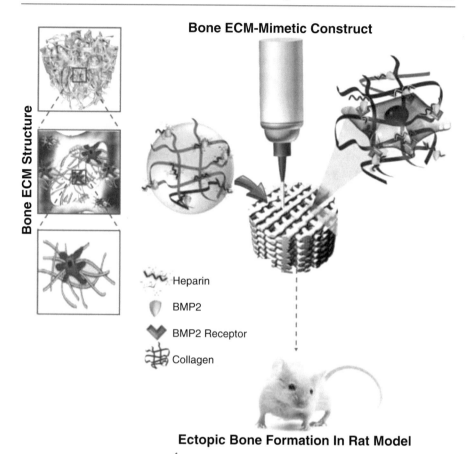

Fig. 6.4 The hybrid structure of the biomimetic constructs with potential applications in personalized bone regeneration. MSCs are encapsulated within collagen–heparin hydrogels functionalized with BMP2. The cell-laden hydrogels are injected into the 3D printed ceramic scaffolds. Reproduced from [82] with permission

were encapsulated within the functionalized hydrogels. Finally, the hydrogels encapsulating the MSCs were injected into the 3D printed porous β-TCP constructs. Disk-shaped MSC-laden scaffolds with 7-mm diameter and 2-mm thickness were implanted at the subcutaneous dorsum of male Fischer rats for 6 weeks. The results showed that due to their excellent biocompatibility, heparin-functionalized collagen hydrogels injected into the 3D printed β-TCP constructs retain the bioactivity of rhBMP2 and support the survival and differentiation of MSCs. The biomimetic constructs developed in this study provided a favorable microenvironment for the growth and differentiation of osteogenic MSCs. The implanted constructs also induced ectopic osteogenesis in rats. The findings of the study provided a practical strategy to improve the efficiency of cell and growth factor delivery and can be applied in hard tissue regeneration.

Several natural and synthetic biomaterials are used to print hard tissue substitutes. The most commonly used biomaterials for this purpose include bioglass, hydroxyapatite, calcium phosphate, collagen, PCL, PLGA, and titanium [83]. These biocompatible biomaterials are printed into osteoconductive scaffolds with desired shape and dimensions. 3D printing facilitates the control of scaffold's porosity and pore connectivity.

Hard tissue regeneration requires supraphysiological concentrations of osteogenic growth factors. Controlled release of the growth factors delivered via the scaffold promotes bone formation and angiogenesis. 3D printing allows for site-specific delivery of drugs and growth factors [83, 84].

In 2019, Cheng et al. [85] used 3D printing to fabricate three-dimensional PLA scaffolds with an interconnected microporous structure applicable in hard tissue engineering (Fig. 6.5). The scaffolds were fabricated by fused deposition modeling (FDM). Then, they were grafted with human BMP-2. Human MSCs (hMSCs) were cultured within the scaffolds, and the osteogenic activities of the cells were evaluated to assess the efficacy of the printed scaffolds. Furthermore, the release behavior of BMP-2 was studied for 35 days. The results showed the higher ALP activity and osteocalcin expression for hMSCs cultured on BMP-2 coated scaffolds, compared to uncoated scaffolds. The study suggested that 3D printed scaffolds coated with growth factors possess favorable physiochemical and biological properties and could be used for bone tissue engineering.

In another study, Kumar et al. [86] fabricated a biphasic construct with potential applications in vertical bone regeneration. The bone-mimicking construct was composed of a 3D printed PCL scaffold loaded with hyaluronic acid (HA) hydrogel. PCL scaffolds were fabricated via FDM to provide the mechanical stability and

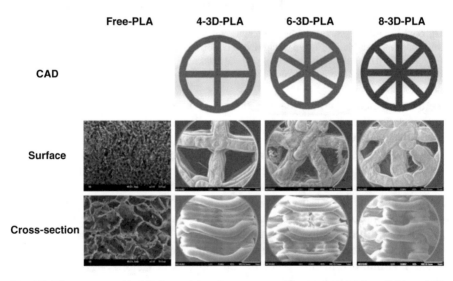

Fig. 6.5 The computer-aided design and the microstructure of freeze-dried PLA scaffolds and 3D printed PLA scaffolds with different morphologies. Reproduced from [85] with permission

volume preservation required for bone regeneration. The HA hydrogel was used as the carrier for the osteogenic growth factor BMP-2. To perform in vitro studies, disk-like PCL scaffolds with 5 mm diameter and 1 mm thickness were fabricated and subsequently loaded with BMP-2 and human primary osteoblast (HOB) cells. Results showed the sustained release of BMP-2 for 28 days. In addition, the viability of HOB cells was well maintained for 3 weeks. qRT-PCR data showed the improved expression of osteogenic markers including osteopontin, osteocalcin, and collagen 1A1 for 2 weeks. The study suggests that bone-mimicking biphasic constructs could be considered as promising scaffolds for bone regeneration and preserving bone height after placing dental implants.

Different materials and particles have been included in the 3D printed scaffolds to enhance their hard tissue regeneration and mechanical properties. For example, in a recent study on 3D printed beta-tricalcium phosphate (β-TCP)-based scaffolds, functionalized reduced gelatin/graphene oxide– magnesium–arginine have been included in the scaffolds (Fig. 6.6) [87]. It was shown that this incorporation could improve the osteogenic capacity of DPSCs. Moreover, the mechanical properties of the scaffolds

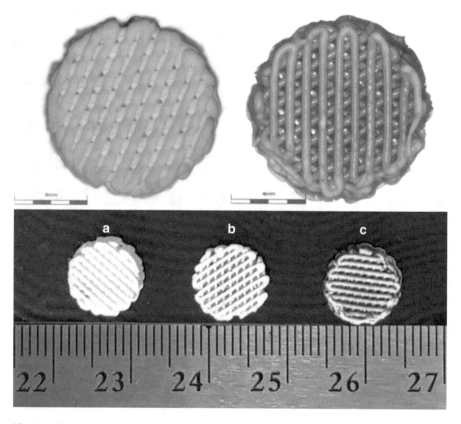

Fig. 6.6 3D printed β-TCP scaffolds functionalized with different concentrations of reduced gelatin/graphene oxide–magnesium–arginine. Reproduced from [87] with permission

were enhanced by the inclusion of reduced gelatin/graphene oxide–magnesium–arginine. The authors optimized the amount of graphene oxide–magnesium–arginine in the composition to get the best biological and mechanical properties [87].

In another study, TiO_2 was included in the 3D printed PLGA scaffold to induce favorable physical and biological characteristics to the scaffold (Fig. 6.7) [88].

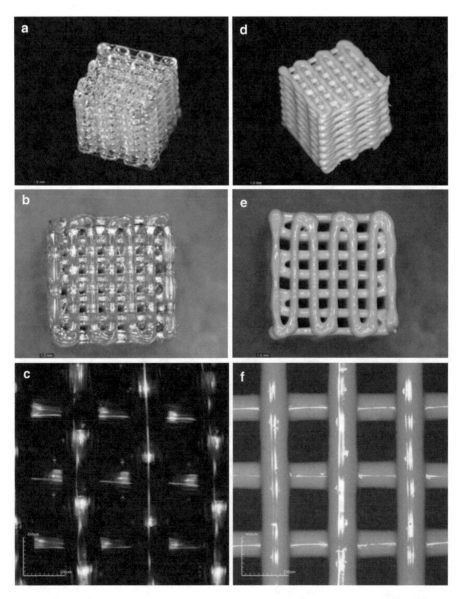

Fig. 6.7 3D printed scaffolds made with PLGA (**a–c**) and composites of PLGA/TiO_2 (**d–f**). Reproduced from [88] with permission

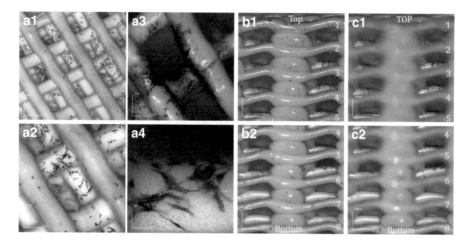

Fig. 6.8 Attachment and migration of osteoblast cells in 3D printed PLGA/TiO$_2$ scaffolds. (**a1–a4**): Images of the cell/scaffold construct in different magnification. (**b1–c2**) Images of cells penetrated into different depths at different focal planes of confocal laser microscopy. Reproduced from [88] with permission

More specifically, by combining PLGA and TiO$_2$ with the weight ratio of 10:1 (PLGA:TiO$_2$), the glass transition temperature and thermal decomposition onset were increased, and the compressive modulus of the scaffold made by this composition was enhanced [88]. While the scaffold made by pure PLGA is hydrophobic, which can decrease the cell attachment, the wettability of the scaffold is increased by the addition of TiO$_2$ to the composition of the scaffold, and thus osteoblasts seeded in the PLGA/TiO$_2$ scaffold presented better proliferation, migration, alkaline phosphatase (ALP) activity, and improved calcium secretion [88]. The migration and penetration of osteoblast cells into the 3D printed PLGA/TiO$_2$ scaffold can be seen in Fig. 6.8.

The property of 3D printed scaffolds not only can be improved by the addition of specific particles with different properties into the main structure of the scaffold but the addition of a separate phase to the construct can be beneficial too. For example, Fahimipour et al. [89] included a collagenous matrix embedded in the 3D printed β-TCP scaffold. As shown in Fig. 6.9, in this design, the 3D printed β-TCP scaffold acts as the bone mineral phase, and freeze-dried collagen/TCP matrix acts as the bone ECM phase. This combined 3D printed/freeze-dried scaffold presented improved mechanical properties and enhanced dental pulp cells' proliferation and ALP activity compared to the 3D printed β-TCP scaffold without the ECM phase [89]. The authors suggested this design as a new platform for osteoblastic differentiation of dental pulp cell for regeneration of the craniomaxillofacial bone [89].

Metal 3D printed scaffolds were also fabricated for bone defect treatments. For example, in a recent work by Hindy et al. [90], the SLM printing method was used to fabricate a specific 3D printed construct made of a titanium alloy with a dense

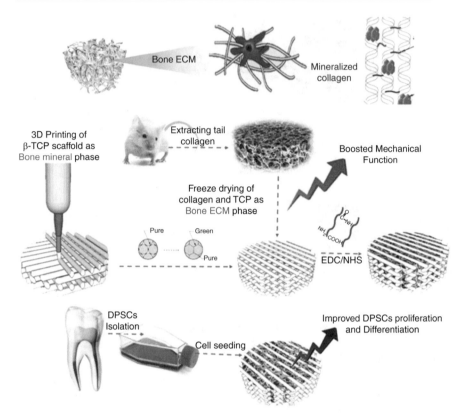

Fig. 6.9 The design of a scaffold with two phases: 3D printed scaffold as the bone mineral phase and freeze-dried collagen/TCP matrix as the bone ECM phase. Due to the enhanced mechanical properties and improved dental pulp cells' proliferation and differentiation (Fig. 6.10), this design can be a new platform for osteoblastic differentiation of dental pulp cells for the regeneration of craniomaxillofacial bone. Reproduced from [89] with permission

core, surrounded by a porous compartment. As can be seen in Fig. 6.11, the porosity of that compartment has a gradient configuration [90].

Since fully dense titanium constructs are mechanically stronger than human bone, the stress shielding phenomena can induce unfavorable effects on the surrounding bone. Making the hybrid dense/porous construct lowers the elastic modulus of the scaffolds to be closer to real bone. Decreased density and enhanced tissue attachment are the other benefits of making such hybrid scaffolds [90].

6.3.3   Bioprinting of Oral and Maxillofacial Hard Tissues

In bioprinting, the cells entrapped in the bioink are extruded through the nozzle to deposit at the desired locations on the substrate. The rheological properties of the bioink are critical to create cell-laden constructs with predefined structure in a

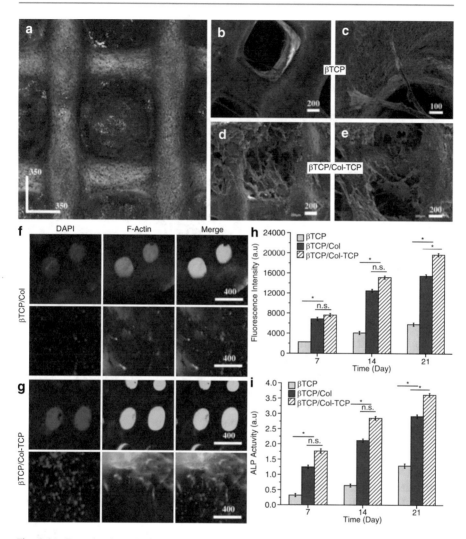

Fig. 6.10 Dental pulp cells grown inside the scaffolds, with the design explained in the previous figure (Col is the abbreviation of collagen). Reproduced from [89] with permission

reproducible manner. For a bioink to be printable, it should possess the thixotropic (shear thinning) property. The thixotropic fluids flow under the shear stress and solidify after extrusion [91].

The main challenge is to obtain a balance between printability and cytocompatibility of the bioink. The low-viscosity bioinks maintain the viability of the cells, but they flatten after being printed on the surface and deviate from the desired structure [92]. To solve this problem and improve the shape fidelity and printing quality, one solution is to blend the bioink with high-viscosity polymers. This should be done carefully because the extrusion of high-viscosity bioinks requires high pressure, which creates high shear forces and affects the viability of the cells [93].

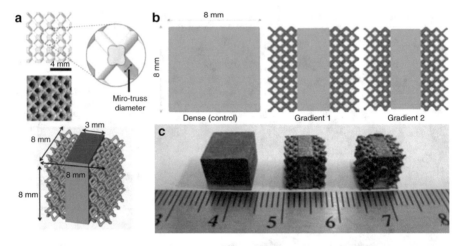

Fig. 6.11 Different schematics (**a**, **b**) and actual images (**c**) of the SLM 3D printed functionally graded porous scaffolds made of a titanium alloy. Reproduced from [90] with permission

The ideal bioink should have proper flow behavior and printing fidelity. It should also provide a tissue-resembling microenvironment to support cellular growth, proliferation, and differentiation [94]. Several studies have been conducted to develop bioinks for printing oral and maxillofacial tissues.

In 2019, Raveendran et al. [94] studied the printability of gelatin methacryloyl (GelMA) hydrogels to fabricate 3D constructs of PDLCs using a microextrusion bioprinter. The goal of the study was to optimize different printing parameters, including GelMA and photoinitiator concentration, printing pressure, printing speed, UV regimen used for photopolymerization, and diameter of the dispensing needle to achieve maximum viability of the printed cells. The optimized bioink (12.5% (w/v) GelMA+0.05% (w/v) photoinitiator) allowed for continuous printing of porous scaffolds with uniform strand diameter and appropriate handling (Fig. 6.12). The cell viability studies showed that the optimized printing parameters did not induce cytotoxicity. The results of the study demonstrated that bioprinting of GelMA containing PDLCs at the optimized conditions results in high-resolution constructs with high cellular viability. The findings of the study could pave the way to utilize PDLCs in the regeneration of periodontal lesions through 3D bioprinting.

6.4   3D Printing in Oral and Maxillofacial Soft Tissue Regeneration

6.4.1   Treatment Options to Reconstruct Oral and Maxillofacial Soft Tissue Defects

The soft tissue specific for oral and maxillofacial region is oral mucosa, which covers the inside of the oral cavity. Three distinct types of oral mucosa include (1) masticatory mucosa, (2) specialized mucosa, and (3) lining mucosa. The

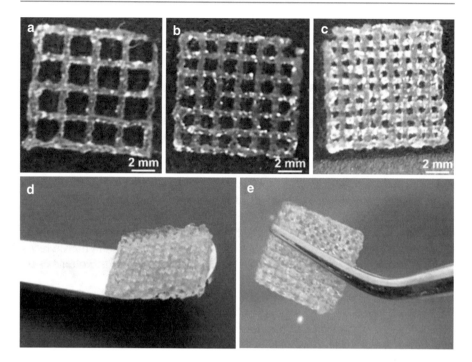

Fig. 6.12 3D printed GelMA hydrogels with different pore sizes: 179.5 ± 9.4 mm (**a**), 94.4 ± 9.3 mm (**b**), and 47.0 ± 8.7 mm (**c**). Handling of the cross-linked hydrogels with conventional surgical instruments (**d**, **e**). Reproduced from reference [94] with permission

masticatory mucosa forms gingiva and covers the hard palate, and the specialized mucosa covers the tongue. The lining mucosa covers the other portions of the oral cavity including the lips, cheeks, and mouth floor [95].

Tumors, trauma, infections, and gingival recessions cause oral mucosa deficiency and necessitate regenerative treatments. The proper treatment is the one that does not interfere with chewing, speaking, esthetics, and other oral functions [68].

To restore oral mucosa defects, autologous free gingival grafts (FGG) and subepithelial connective tissue grafts (SCTG) are commonly used [96]. However, using autologous grafts have several limitations, including:

1. Difficult process of tissue harvesting.
2. Limited volume of the harvested tissue.
3. Dimensional diversity at different parts of the donor tissue.
4. Postsurgical complications including prolonged pain, numbness, and increased risk of infection [97–101].

Soft tissue allografts have been introduced to replace the autografts and substitute oral soft tissues in mucogingival surgeries. One of the first products to treat mucogingival defects was freeze-dried skin (FDS) allografts. FDS allografts were considered as candidates to replace FGGs. The main application of FDS allografts was in root coverage procedures for the augmentation of keratinized gingiva. For this purpose, FGG allografts were used after performing an apically positioned flap [68].

Another substitute for oral soft tissues was the acellular dermal matrix (ADM) allograft. ADM allografts were initially developed in the 1980s for covering full-thickness burn wounds. However, they found several applications in mucogingival surgeries including the augmenting of keratinized gingiva and localized alveolar defects, covering of exposed dental roots, and increasing the depth of the vestibular fornix [68, 102]. The applications of ADM allografts are limited due to the high shrinkage and difficult handling. Moreover, clinical studies have revealed substantial differences between the grafted ADMs and natural tissues [103, 104].

Mucograft is a novel xenogeneic resorbable collagen matrix developed by Geistlich Pharma to boost the healing process and reduce shrinkage. It is obtained from porcine collagen (type I and III) with no chemical treatments employed during the processing. Mucograft has been clinically used as a coverage for gingival recessions. Clinical studies demonstrate that it successfully increases the volume of the keratinized gingiva [105, 106].

These novel scaffolds developed from allografts and xenografts eliminate the problems of harvesting and morbidity. They also decrease operational time and costs. However, they need tailoring for each defect and do not imitate the natural structure of the oral and maxillofacial tissues. Besides, high shrinkage and difficult clinical handling limit the application of allografts and xenografts [68, 104].

6.4.2   Requirements and Advances in 3D Printing of Oral and Maxillofacial Soft Tissues

To overcome the limitations of tissue grafts, soft tissue substitutes are required [107, 108]. The essential requirements for an ideal soft tissue substitute are biocompatibility, biodegradability, tissue integration, mechanical and dimensional stability, low cost, and easy handling [107].

Oral and maxillofacial soft tissue reconstruction is among the most challenging and demanding medical conditions and mainly involves epithelial and connective tissue treatments. Soft tissue repair is an indispensable part of treatments for many oral and dental complications such as gingival recession and tooth loss. The most frequent causes for oral and maxillofacial soft tissue lesions are traumas, infection, inflammation, and tumors [109, 110]. Surgical treatment is needed to treat most of these lesions. The treatment should meet the terms of functional mastication, phonetics, and aesthetics. The surgeries are commonly performed using conventional and electric scalpel and lasers. In almost all surgical procedures, surgical or thermal damage of the peripheral healthy tissues is almost inevitable. For this reason, and also because of the possibility of lesion recurrence after surgical treatment, regeneration of the soft tissue lesions seems to be a safer and more effective treatment option.

One of the most prevailing oral soft tissue complications is gingival recession, in which the gingival margins move apically from the cementoenamel junction (CEJ) and result in root exposure. The most common consequences of root exposure are dental hypersensitivity, phonetic impairments, and aesthetic problems [111].

Treatment of gingival recession is important to improve the aesthetic quality and decrease dental sensitivity. In addition, the recession should be controlled, unless it spontaneously progresses, even in patients with good oral hygiene [112]. The ideal treatment outcome is the complete recovery of the teeth roots. The availability of interdental papilla plays the most critical role to obtain the best results of root coverage [113].

The interdental papilla is a connective tissue, composed mainly of dense epithelium, and fills the gap between the adjacent teeth. The loss of interdental papilla has been demonstrated to cause problems for patient's chewing, speaking, and aesthetics. Regeneration of papilla is one of the most complicated periodontal procedures. Several surgical and nonsurgical methods are used for papilla reconstruction, including interproximal tunneling, patch technique, hyaluronic acid injection, hyaluronic acid overlay technique, platelet-rich fibrin (PRF), and interposed connective tissue grafting. However, there is still no reliable technique for this purpose [114–123].

Recently, 3D printing has been considered as a potential method for the treatment of soft tissue lesions. For example, a 3D printed gradient gelatin membrane was suggested to be used for the reconstruction of full-thickness oral mucosa [124]. In a study by Stuani et al. [125], after reviewing the current therapeutic choices, 3D printed scaffolds were proposed for the treatment of gingival recession.

6.4.3   Bioprinting of Oral and Maxillofacial Soft Tissues

In contrast to 3D printed scaffolds, which have found increasing applications, the use of bioprinted constructs in the regeneration of oral and maxillofacial tissues is still in its early stages. Until now, 3D bioprinting has been successfully applied for the regeneration of other similar tissues, such as skin and vascular tissue constructs. The knowledge obtained from these studies may provide practical cues to fabricate 3D printed, patient-specific oral and maxillofacial soft tissues with satisfying functional and aesthetic results [126].

Based on the dimensions of the defects, specific considerations should be given to successfully regenerate oral and maxillofacial hard and soft tissues. Printing grafts with appropriate shape and architecture could regenerate small-sized defects. The ingrowth of the patient's cells into the implanted grafts provides regeneration mechanisms. For larger defects, autologous cells should be co-printed with the scaffold. The most important consideration in the regeneration of both defect types is the potential of the (bio)printed construct to support vascularization. If the strategy of choice is based on the bioprinting of prevascularized grafts, the hierarchical structure of the vasculature, including arterioles, venules, and capillaries, should be considered [126].

Essential requirements to fabricate gingival-like tissue are depicted in Fig. 6.13. A mix of smart bioinks resembling tissue-specific extracellular matrix components

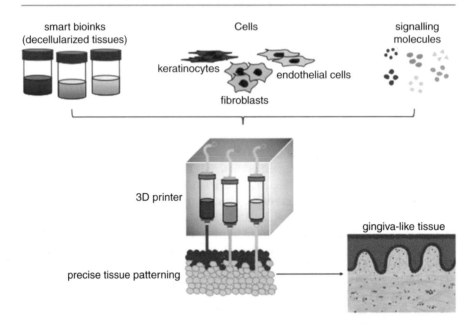

Fig. 6.13 Essential requirements to fabricate gingiva-like tissue constructs. Reproduced from [126] which is distributed under a creative commons license

is required. The bioinks should be co-printed with gingival fibroblasts, keratinocytes, and endothelial cells. Growth factors can also be printed within the constructs as long as they do not lose their bioactivity. The mixture of these constituents should be accurately patterned via inkjet, extrusion, or laser-assisted bioprinting to imitate the complex structure of the gingiva [126].

6.5   Reconstruction of Critical-Sized Defects in Oral and Maxillofacial Surgery

Acute or refractory dental, oral, and maxillofacial defects and disorders present surgeons with numerous reconstructive challenges.

Oral and maxillofacial defects are currently reconstructed using microvascular-free flaps, nonvascular autogenous bone and/or skin grafts, or are reconstructed with xenografts or synthetic materials. Using these methods, surgeons are constrained by the size and complexity of the defect; defects larger than 2.5–5 cm are difficult to reconstruct with these techniques and are those defects that involve the anterior mandibular region. Lack of predictability of graft success, as well as the lack of predictability of the esthetic and functional outcome of a nonvascular grafting, makes it necessary to use a vascularized free flap in many clinical cases with large size and multi-tissue damage. Such procedures are lengthy and challenging despite improvements in technology and techniques. Patients experience long recovery periods and are exposed to related perioperative morbidities that increase healthcare costs and time lost from work, and negatively affect the patients' quality of life.

Spontaneous renewal of the bone is limited to small defects and critical-sized defects caused by tumors, trauma, implant loosening, or osteitis requiring specific handling or surgical therapy. A critical-sized defect in treatment can be defined as a size greater than which the defect cannot heal and regenerate by itself without planned reconstruction and regeneration therapy [127]. The exact size of a bony defect to be considered as a critical-sized defect is not clear, but since 2.5 cm bone defects showed poor natural history, it is reasonable to consider that as a critical value for the size of the defect [127].

Please note that nonunion bone defects are different from critical-sized defects. The biomechanical problems and cellular/molecular signaling impairment are the usual characteristics of nonunion bone defects, while there might be no bone gap in the structure of the defect. However, in critical-sized defects, the biology and signaling can be acceptable, but the large area of the defect which can be accompanied by soft tissue loss makes tissue regeneration difficult [128].

Different methods have been utilized to meet the huge demand of critical-sized bone healing. Autografts are the gold-standard treatment for repairing bone injuries. The major problem with this approach is the limited supply and donor site morbidity. Alternatively, bone allografts can be used instead of autografts. However, there are serious cost issues and a risk of disease transmission in using bone allografts. Metallic implants are also widely used in bone treatment. However, unlike natural bone, they are not a self-repairing material. Also, they do not adjust with alterations in physiological conditions. More importantly, the implants can become movable and loose over time.

The deficiencies of the abovementioned approaches have made bone tissue engineering popular during the last decade. There are some obstacles and challenges in applying the tissue engineering approach for the treatment of critical-sized oral and maxillofacial defects, including the lack of enough mechanical strength of conventional scaffolds, vascularization, and the healing time [129].

6.5.1   Mechanical Strength of Scaffolds

As we discussed before, one of the most important obligations of tissue engineers is developing scaffolds. Generally, scaffolds in tissue engineering are the initial biomechanical profile to organize the seeded cells until they produce an adequate extracellular matrix and develop into the desired tissue. Therefore, tissue scaffolds must present sufficient mechanical properties matched to the site of implantation, along with controlled biodegradability for being replaced by tissue. A good scaffold is biodegradable with a porous structure. However, porosity and degradation of the scaffolds inversely affect their mechanical properties. For the bone scaffolds to be used in clinics, it is expected to have sufficient mechanical strength to bear the load of the body while facilitating bone regeneration.

The ultimate goal in designing scaffolds for bone tissue engineering is to mimic the morphology, structure, and function of bone. Currently, different kinds of materials have been utilized for this purpose, which is synthetic or naturally derived.

To repair bone defects, 3D porous scaffolds provide more advantages than the normal powder or granules. Ideal porous scaffolds consist of interconnected macroporous networks allowing cell migration, nutrient delivery, and eventually vascularization.

However, the porosity significantly decreases the mechanical strength of the scaffolds, which is crucial for critical-sized defects.

Different synthetic and natural polymers with various fillers have been employed to obtain desired mechanical properties for bone tissue engineering scaffolds as described in Table 6.1 [130]. With a quick glance at this table and comparing the mechanical properties of the obtained scaffold with the mechanical properties of human bone tissue, it helps us to realize the depth of the problem.

The trade-off between sufficient mechanical properties and porosity is one of the major challenges in designing tissue-engineered constructs. Ikida et al. [138] measured the modulus of the PLG scaffold with different porosities at 10% strain. They observed that with increasing porosity from 80% to 92%, modulus dramatically decreases from 0.26 MPa to 0.0047 MPa (~50 times) [138]. Furthermore, in current investigations, scaffolds are often tested mechanically in their dry state; thus, the effect of the aqueous medium of the body is neglected. Many biocompatible and biodegradable polymers, such as gelatin, collagen, or chitosan, which are necessary for fabricating tissue scaffolds, considerably swell in aqueous mediums. In this case, modulus will be inversely related to water content [139, 140]. Swelling may cause a remarkable decrease in modulus (~3 times), even in low water uptake (<5%) [140]. Scaffolds made from these polymers represent better mechanical properties (modulus in the range of 1–100 MPa) when tested in a dry state, which does not mimic the environment of the human body. When the scaffolds fabricated from these materials are tested in the swelled state, the modulus drops significantly (to less than 100 kPa) [141–143].

Table 6.1 Mechanical properties of human bone tissue and different materials employed for the preparation of bone tissue scaffolds [130]

Scaffold components		Mechanical properties with porosity			
Polymer	Filler	Modulus (MPa)	Strength (MPa)	Porosity (%)	Ref.
PLLA	–	2.2	1.8	87	[131]
PLLA	Nano-HA	14.9	8.7	85	
PLGA(50:50)	–	8.33	1.2	92	[132]
PLGA(50:50)	40%CPC	–	8.2	–	[133]
PCL	–	–	0.27	–	[134]
PCL	HA	–	0.57	–	
Chitosan	–	4.5	–	–	[135]
Chitosan	2% Nano-HA	8.6	–	–	
Gelatin	50% BaG	78	5.6	72	[136]
Cancellous bone		20–500	4–12	50–90	[137]
Cortical bone		3000–30,000	130–180	5–10	

To address the issue of mechanical strength and porosity, we manufactured a novel-designed hybrid scaffold with significantly improved mechanical properties, not only at the dry state but also at the simulated environment of a human body [130]. In this design, a skeleton with strong mechanical properties is employed to be responsible for the load-bearing, and a component with good bioactivity is molded into this skeleton to be responsible for the biological properties [130]. More specifically, the design is inspired by reinforced concrete, in which metal reinforcing bars (rebars) are embedded passively in the concrete; thus, the concrete's relatively low strength is counteracted by the inclusion of reinforcement (Fig. 6.14a). A regular bone tissue engineering scaffold, as can be seen in Fig. 6.14b, is composed of porous and degradable materials, so the cells can grow inside the pores, and the scaffold itself degrades in the body after the new bone tissue is formed. In the proposed design, as can be seen in Fig. 6.14c, the scaffold is composed of two specific components: (1) Skeleton: non-porous and slowly-biodegradable constituent to keep the mechanical integrity of the scaffolds for long period and (2) Host Component: porous and rapidly biodegradable constituent to allow the cells grow inside while it will be degraded [130]. The skeleton is placed into the Host Component to reinforce it (Fig. 6.14c).

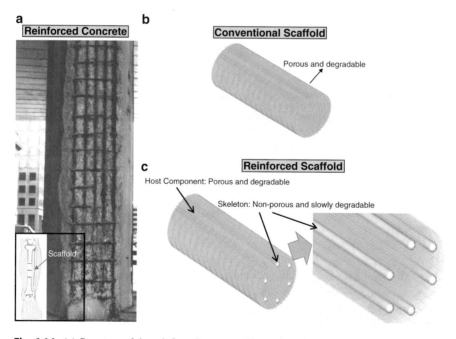

Fig. 6.14 (**a**) Structure of the reinforced concrete. The probability of structure failure is significantly low due to the embedded rebars with strong mechanical properties. (**b**) Structure of a conventional bone scaffold. Although the porosity is required to make an appropriate environment for cell growth inside the scaffold, it significantly decreases the mechanical properties. (**c**) Inspired by the construction of reinforced concrete, a new scaffold can be made which is composed of a Skeleton with high relative strength, embedded in a porous degradable material. This significantly enhances the mechanical properties of the scaffold so it can bear the load of the human body. This kind of scaffold can be specifically used for critical-sized defects

When the scaffold is implanted in the body, bone cells start growing into the Host Component. Due to the porosity of this part, cells can penetrate inside. The Host Component is responsible for cell regeneration and attachment. While the bone cells proliferate, the host component will be degraded over time.

The Skeleton of the scaffold is responsible for its mechanical strength. To be more specific, the material selected for the Skeleton should have the following properties: (1) must maintain its mechanical properties in human body condition, (2) must support the defect mechanically until proper formation of new tissue (mechanical properties should not drop dramatically at the first stages of tissue formation); (3) its degradation products must be biocompatible; and (4) its degradation rate and mechanical properties should be adjustable for long-term healing procedures.

The materials selected for the Host Component should be: (1) biocompatible with high cell attachment capability, (2) degradable in the short term, and (3) capable to be fabricated in a porous form.

As can be seen, both Skeleton and Host Component should be biodegradable, but the degradation rate should be faster for the Host Component. Both could be fabricated using pure or composite materials including biodegradable polymers, ceramics, or even metals.

To summarize, Table 6.2 lists the requirements of each part of the scaffolds.

Since the porosity of materials decreases the mechanical strength of the Skeleton, it should be dense (non-porous), made with strong materials, and have slow biodegradability. Some suggested materials to be included in the composites of the skeleton includes:

1. Poly(α-esters) (e.g., Polyglycolide, Polylactides, Poly(lactide-co-glycolide), Polydioxanone, Polycaprolactone, Poly(trimethylene carbonate), and Bacterial polyesters).

   Aliphatic polyesters with reasonably short aliphatic chains between ester bonds can degrade over the time frame required for most of the biomedical applications. Poly(α-esters) comprise the earliest and most extensively investigated class of biodegradable polymers. They are attracting significant attention as biomaterials due to their good biocompatibility and controllable degradation profiles.
2. Polyurethanes.
3. Poly(ester amide).

   Due to the hydrogen bonding ability of the amide bonds and biodegradability imparted by the ester bonds, these copolymers have good mechanical and thermal properties. The degradation of poly(ester amides) has been shown to

Table 6.2 Requirements for each part of the scaffold

Requirement	Skeleton	Host component
Good mechanical properties	Required	Not required
Porosity	Not required	Required
Biocompatibility	Required	Required
Controlled biodegradability	Required	Required

take place by the hydrolytic cleavage of the ester bonds, leaving the amide segments more or less intact.

4. Poly(orthoesters).

   Poly(orthoesters) were developed by the ALZA corporation (Alzamers) as a hydrophobic, surface eroding polymer. Although the orthoester linkages are hydrolytically labile, the polymer is hydrophobic enough such that its erosion in aqueous environments is very slow. The unique feature of poly (orthoesters) is that, in addition to its surface erosion mechanism, the rate of degradation for these polymers, pH sensitivity, and glass transition temperatures can be controlled by using diols with varying levels of chain flexibility.

5. Polyanhydrides.

   Polyanhydrides can be considered as the most extensively investigated biodegradable surface-eroding polymers. The hydrolytically labile backbone coupled with the hydrophobicity of the polymer precludes water penetration into the matrix, allowing polyanhydrides to truly undergo surface erosion.

6. Poly(anhydride-co-imide).
7. Cross-linked polyanhydrides.
8. Poly(propylene fumarate).
9. Pseudo poly(amino acid).
10. Poly(alkyl cyanoacrylates).
11. Polyphosphazenes.
12. Polyphosphoest.

Biocompatible hydrogels, in porous configurations, can be good candidates for the Host Component to provide an appropriate medium for cells to attach, grow, and proliferate. Some examples include:

1. Proteins and Poly (amino acids) (e.g., Collagen, Gelatin, Natural poly (amino acids), Synthetic poly (amino acids), Elastin, Elastin-like peptides, Albumin, and Fibrin).
2. Polysaccharides (e.g., Polysaccharides of human origin, Polysaccharides of non-human origin).

Different preparation methods can be employed for the production of the Skeleton and Host Components. Although 3D printing is very appropriate to be used for preparing the Skeleton (Fig. 6.16), molding is also possible to be used (Fig. 6.15). To prepare the porous Host Component, freeze-drying is a practical approach; however, one may use other methods, such as 3D printing, solvent casting/particulate leaching, gas foaming, porogen leaching, self-assembly, phase separation, melt molding, fiber bonding, and membrane lamination.

As shown in Fig. 6.16a1–a3, the reinforcement could be performed in one, two, or all three directions (x, y, z). Skeleton can be assembled in various configurations as shown in Fig. 6.16b1–b3, such as box-shaped, cylinder-shaped, and honeycomb-shaped.

Scaffolds will have different mechanical properties based on the configuration and material of the Skeleton. An example of the mechanical properties of one prepared scaffold can be seen in Fig. 6.17.

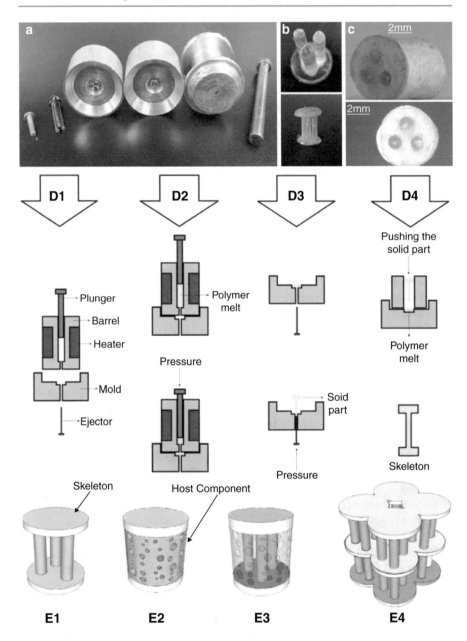

Fig. 6.15 (**a**) Molding instruments (**b**) Produced skeletons (**c**) Fabricated hybrid reinforced scaffolds, (**d1–d4**) Manufacturing procedure of skeleton. (**e1–e3**) Inclusion of Host Component to the Skeleton. (**e4**) Enlarging the scaffold by joining some Skeleton structures [130, 144]

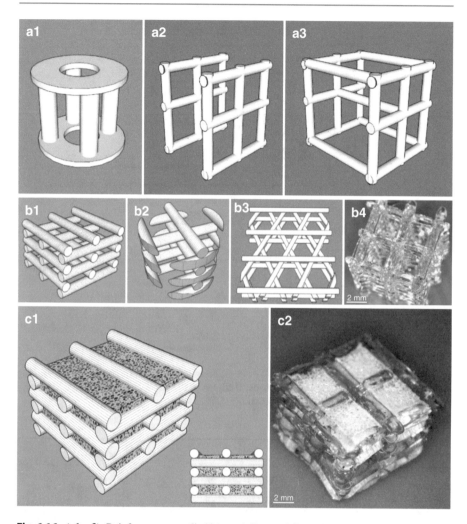

Fig. 6.16 (**a1**–**a3**): Reinforcement applied in one (**a1**), two (**a2**), or three directions (**a3**). (**b1**–**b3**) Examples of various assemblies of skeleton: (**b1**) Box-shaped, (**b2**) Cylinder-shaped, (**b3**) Honeycomb-shaped, (**b4**) The actual 3D printed box-shaped skeleton. (**c1**) Host component will then be incorporated into the skeleton. (**c2**) Prepared scaffold composed of both skeleton and host component [130]

The exact configuration of the Skeleton to have an optimized property of the reinforced scaffold can be different for various defect shapes and the area of injury in the oral and maxillofacial region based on the stress distribution on that area and the surrounding tissue. Please note that the mandible mechanical performance and behavior are different in different areas (the exact location in which the scaffolds must be implanted).

To figure out such configuration, as can be seen, an example in Fig. 6.18, one may start from the CT (3D computed tomography) scan image of the patient's injured mandible area, locate the critical-sized defect, and extract the 3D STL file

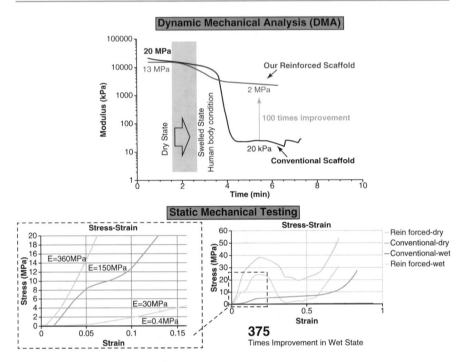

Fig. 6.17 Mechanical properties of one hybrid reinforced scaffold and static mechanical testing show that our reinforced design improves the modulus of the scaffold 100 times and 375 times, respectively, in a simulated human body condition [130, 144]

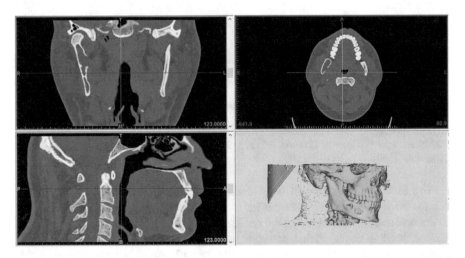

Fig. 6.18 Stack-up of the 3D CT images: the procedure starts with the CT image analysis of the defect area

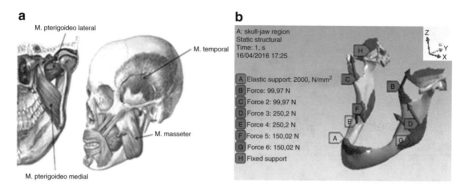

Fig. 6.19 Panel (**a**) shows the major mastication muscles and panel (**b**) represents the boundary conditions that be used in the modeling. Reproduced from [150] which is distributed under the Creative Commons Attribution License

from the DICOM (Digital Imaging and Communications in Medicine) image. For example, assume that the critical-sized defect is in the area shown in Fig. 6.20a.

To determine the applied mechanical load on the scaffolds, the 3D finite element model (FEM) of the mandible with the critical-size defect using the solid tetrahedral element can be developed (the STL file can be imported to the FEM for finite element analysis).

Literature can guide us to find the boundary conditions of the model, characteristics of the material, and muscle forces affected in the model [145–150].

For example, Fig. 6.19a displays the muscle arrangement, and Fig. 6.19b can help in figuring out the applied boundary conditions, positions, and orientations of muscle forces, and constraints considered for the model [150].

Using the geometry and materials properties, jaw anatomy, and FEM stress analysis, loads applied to the main muscles of mastication and boundary conditions as well as the stress analysis of the mandible and the critical defect, scaffold mass optimization can be performed. For example, this optimization for the defect shown in Fig. 6.20a, b is illustrated in Fig. 6.20c which is based on the geometry optimization considering the minimum strain applied to the internal area, with maximum load-bearing capability and using minimum amounts of materials. The bridge shape shown in Fig. 6.20d, e is basically the shape of the found Skeleton with maximum possible mechanical strength using the minimum amount of nonporous material that can be used for the reinforced scaffold. The Host Component then will be incorporated into such a Skeleton to make the optimized hybrid reinforced scaffold for the defect shown in Fig. 6.20a.

6.5.2 Vascularization

As the poor mechanical properties of current scaffolds prevent commercial applications of such products, this design can pave the way to the commercialization of tissue engineering bone scaffolds, especially for load-bearing applications. However,

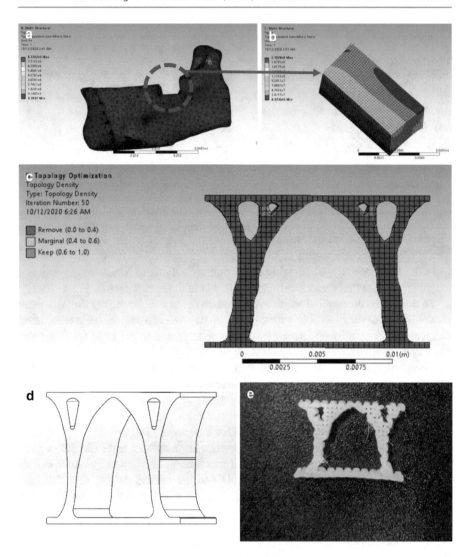

Fig. 6.20 Assuming that the defect of the mandible has the shape and location shown in panels (**a**) and (**b**), the bridge shape illustrated in panels (**c**) and (**d**) are the optimized shape of the Skeleton for the reinforced scaffold that should be used for the treatment of such defect. Panel (**d**) is the actual 3D printed structure of the Skeleton. To make the final reinforced scaffolds, the host component must be incorporated into the Skeleton [151]

the mechanical strength of the scaffolds is not the only challenge in the handling of critical-sized defects.

There are specific clinical problems associated with poor vascularity in critical-sized craniomaxillofacial defects, in which the size of the defect hinders vascularity [152–155] or even in the small defects that have to be treated in special conditions (such as after radiation therapy, e.g. after tumor resection) that create poor vascularity [155–159]. Moreover, in methods of seeding stem cells into the graft materials for bone defects, the cell viability in the deep portion of scaffolds is compromised as a result of poor vascularization [160–162]. In general, problems in vascularization lead to a lack of graft integration and the ultimate failure in the treatment of craniomaxillofacial defects.

There are five strategies including (1) growth factor delivery such as the release of vasculogenic and angiogenic agents in the area of defects; (2) co-culturing systems and use of endothelial cells such as coculturing of human osteoblast-like cells (HOBs) with human umbilical vein endothelial cells (HUVECs); (3) mechanical stimulation as external mechanical stimuli can enhance vascularization in bone tissue engineering approaches; (4) the use of biomaterials with the requisite properties; and (5) incorporation of micro-fabrication techniques for advanced control over the microenvironment [163, 164].

Among the abovementioned strategies, growth factor delivery by the means of a vehicle (e.g., particle) has proved to be the most effective approach [163, 165, 166]. Vascular endothelial growth factor (VEGF) is a vital angiogenic GF that can be used in combination with other agents, such as BMP-2, bFGF, and KGF, to enhance bone regeneration [167–172]. For example, it has been shown that the mineral density of de novo bone is increased by 33% in bone scaffolds with both BMP-2 and VEGF, when compared to those with BMP-2 only [170].

VEGF controls the osteoblast differentiation, function, and coupling of osteogenesis to angiogenesis [173]. An investigation in mice showed that the removal of VEGF causes a major decrease in the recruitment of macrophages, stimulation of angiogenesis, and the density of mesenchymal precursor at the injury site, which in turn leads to a significant delay in bone regeneration [174]. The coupling of angiogenesis and osteogenesis is regulated through different mechanisms by VEGF [175–177]: (1) VEGF directly affects the osteogenic differentiation of mesenchymal stem cells (MSCs) [178]. Here, VEGF stimulates osteoblastic differentiation and inhibits adipogenic differentiation of MSCs by intracellular signaling using transcription factors RUNX2 and PPARy2 [179, 180]. (2) VEGF induces angiogenesis and osteogenesis through stimulation of the endothelial cells to express some factors such as BMP, which in turn induces osteogenic differentiation and matrix mineralization [181]. (3) Non-physiological doses of VEGF recruit immune cells and osteoclast precursors and therefore impair bone regeneration and increase bone resorption [174, 182].

The delivery of VEGF is highly challenging, as imprecise amounts of exposure can cause improper neovascularization and carry the risk of tumorigenesis [183]. On the other hand, due to its short half-life, slow and short-lived exposure to VEGF produces transient and often short-lived neo angiogenesis. It has been shown that high local concentrations of VEGF induce the formation of malformed, nonfunctional vessels or even increase bone resorption [182, 184]. Therefore, to avoid such problems, VEGF levels should be tightly managed. Both localized and controlled low dosage sustained release of VEGF are extremely important.

There are different types of carriers that have been designed and offered for GF deliveries, but a vigorous predictive technique for the production of these carriers is yet to be accomplished [84, 167, 185–194]. The current technique for the production of particles as growth factor carriers is the bulk mixing method. This approach is not favorable, as it results in polydisperse particles with different physicochemical characteristics that fail to provide optimal spatiotemporal release. Thus, having the particle of interest is complicated, costly, and often not precise, and post-processing steps are required. Moreover, the bulk method involves harsh mixing steps, which may cause denaturation and reduced bioactivity of the growth factor. In fact, size polydispersity in bulk methods causes the formation of particles with various release profiles, since the size and microstructure directly affect the evaporation rate of solvent in the course of the drying step in the production process and result in undesirable rate of particle degradation, stability, and the kinetics of release [195]. Hence, the current methods cannot fulfill precise temporal and spatial delivery of the VEGF to the microenvironment. Please note that the polydisperse particles may have reasonable accumulative release rates (for all particles together); however, due to the variations in physicochemical properties of particles, each of them may have slightly different release profiles, which is unfavorable and unsafe for VEGF delivery [196, 197]. The release of VEGF must be sustained and localized in the target zone. Uncontrolled release of VEGF can enhance the risk of tumor growth in distant areas [197].

The design of the hybrid reinforced scaffold can be used as a platform for immobilizing biological agents, such as growth factors, and more specifically VEGF, on the surface of the Skeleton of the scaffolds. However, first monodisperse drug-loaded particles with robust release profiles must be fabricated. The fabrication of a VEGF-loaded particle with precise properties might only be possible with the assistance of microfluidic techniques.

Also, depending on the material's type of Skeleton and also the VEGF-loaded particles, a practical approach must be found to firmly immobilize the particle on the Skeleton to facilitate the localized delivery of VEGF.

6.5.3   Healing Time and External Stimulation

The healing time is crucial for critical-sized defects. The period of large bone defect healing can be from several weeks to 2 years. Moreover, when the healing time is extended, there is a risk that the fractures will not join properly. Due to the high demand for decreasing the duration of fracture healing and enhancing the quality of bone healing and fracture fusions, several strategies have been developed such as employing electromagnetic stimuli.

It is known that electromagnetic stimulation has a progressive influence on the fast healing of bone. The first animal tissue response to electrical stimulation was investigated in the eighteenth century and studied by physiologist Galvani who observed the contraction of a frog's muscles when a lighting-rod voltage was applied [198]. This famous experiment demonstrated "animal electricity", which paved the way for the later invention of electric cells and set the foundation of electrophysiology as a new science. For the first time, work on the effect of electrical stimulation on bone tissue was carried out by Yasuda from Japan in 1950 [199]. Yasuda's pioneering studies opened the door for bone healing treatments using electrical stimulation. The animal bone-associated cellular response to the electric field was found in the late 1960s using a dog model [200]. The experiment showed faster bone fracture healing when an electric field was applied. Subsequent experiments on animals proved that the local application of electrical stimulation can improve bone growth [201, 202]. The cellular and molecular mechanisms of enhanced bone regeneration have been studied in the last two decades. Current research combining molecular cloning and genetic and cellular visualization techniques has provided many significant insights into the cellular response to electrical stimulation. From the molecular biological point of view, the straightforward evidence is the gene expressions at different stages of osteoblastic development [203–206]. In brief, the crucial module of neural communication in the bone tissue reacts to the electric stimuli [207].

To combine the technique of employing electromagnetic stimuli into tissue engineering, the conductivity of the 3D scaffolds is required to be adjusted [208]. For example, in the hybrid reinforced scaffolds, with the Skeleton and Host Component, the electrical property of the Host Component can be improved by the addition of conductive polymers. The conductivity of the Host Component will allow the local delivery of electromagnetic stimuli to the site of the defects to enhance the healing procedure. Depending on the composition of the Host Component material, different kinds of conductive polymers can be incorporated into the material. A few suggestions of biocompatible conductive polymers include:

Polypyrrole (PPy): PPy is the most widely held conductive polymer suitable for biomedical purposes. It is highly conductive and has perfect environmental stability. Its preparation method and surface modification are practical and easy. PPy shows good biocompatibility both in vitro and in vivo [209, 210]. It also has a good ion exchange capacity. Its capabilities of adhesion and growth of different cell types were investigated in various reports [211]. Regarding the in vivo investigations of this polymer, the implants of PPy were examined in rat muscle, hypodermis tissue, rat peripheral nerve, guinea pig brain, and peritoneum mice [212–217].

Poly (3,4- ethylenedioxythiophene) (PEDOT): PEDOT has been widely used as conductive coatings for nerve tissue engineering [218]. They are biocompatible with poor biodegradability PEDOT coatings in nerve tissue engineering have been shown to improve the charge transfer characteristics of the underlying surface and cells preferentially adhere to the coated substrate [219–221]. Liver tissue engineering [222] and cardiac tissue engineering [220] have taken advantage of this polymer too. PEDOT has also been used for bone tissue engineering with positive results [219, 221, 223].

Polythiophenes (PTh): The properties of PTh are very comparable to PPy, and in some cases, it is reported to be more promising [224, 225]. However, direct studies on the capabilities of PTh for tissue engineering applications are fewer and relatively newer than PPy and PEDOT [226].

Polyaniline (PAni): Although there are some scattered reports on the capability of PAni for tissue engineering applications, [227] it has not been completely

Fig. 6.21 The idea is to make a device that can induce electromagnetic stimulation (e.g., PEMF) to the area of the critical-sized defect, which is treated by a conductive 3D scaffold. Conductivity of the scaffold facilitates the local delivery of the electrical stimuli to the area of injury to help fast healing of bone. Patients may use such devices daily for a certain period of time (e.g., 30 min/day)

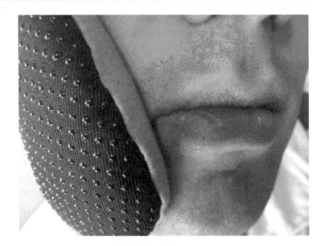

investigated. This polymer has very good conducting characteristics in various structural configurations with acceptable environmental stability. Although based on reports that demonstrate the biocompatibility of PAni and its derivatives both in vitro and in vivo, [228] it is known as a polymer with poor cell compatibility [207]. The in vitro and in vivo biocompatibility of this polymer for bone tissue engineering requires further investigation.

After inducing the conductivity to the scaffold, electromagnetic stimulation can be applied by different methods, such as using the pulsed electromagnetic field (PEMF) as effective stimuli for fast healing of bone fractures (Fig. 6.21) [229].

6.6   Legal and Ethical Issues of Tissue Printing

3D (bio)printing is considered to be one of the major trends of the fourth industrial revolution [230]. Until now, the early development stages of bioprinting technology have been completed. Major advancements have been made in the development of materials and methods of (bio)printing, and the technology has been used to create tissue and organ substitutes. Moreover, the clinical implementability of (bio)printing has been demonstrated [231]. This emerging approach has mainly been used to solve problems in regenerative medicine and transplantology. Although there are still a number of unsolved technical challenges and restrictions, 3D (bio)printing is expected to be a prevailing medical practice in the next two decades. Until then, there is a long way to solve the legal and ethical issues of 3D (bio)printing. The main legal requirements for industrialization and commercialization of 3D (bio) printed products are as follows:

1. Informed consents for donation, manipulation, storage, and further applications of biological sources and tissue derivatives.
2. Development of regulations to guarantee the safety, quality, and efficiency of procedures and end products regarding ethical issues and human rights.

3. Establishment of national rules on technical, legal, and ethical issues about the development and application of 3D (bio)printing technologies.
4. Legislation of mechanisms to ensure that all patients, including minors and incapable people, are protected from the possible adverse effects of (bio)printed products.
5. Establishment of regulations to control the financial turnover of 3D (bio)printed organs and tissues.
6. Obligation to prohibit illegal trafficking of artificial tissues and organs [231–233].

6.7 Summary

3D printing allows for the creation of complex-shaped constructs resembling the natural tissues of the patients. The key strength of this rapid prototyping method is providing a way to efficiently manufacture high volumes of low-cost scaffolds in a reproducible manner. In this chapter, challenges, requirements, and advances for the (bio)printing of oral and maxillofacial hard, soft, and critical-sized defects were discussed. Recent progress in the development of design, materials, and the optimization of (bio)printing techniques promise great advances in realizing the dream of oral and maxillofacial tissue regeneration. However, there is still a long way to go to use 3D (bio)printed tissue and organ substitutes in daily clinical practice. The remaining problems are not just technical as there are many ethical and legal issues about the industrialization and commercialization of 3D (bio)printed products.

References

1. Datta P, Ozbolat V, Ayan B, Dhawan A, Ozbolat IT. Bone tissue bioprinting for craniofacial reconstruction. Biotechnol Bioeng. 2017;114(11):2424–31.
2. Salah M, Tayebi L, Moharamzadeh K, Naini FB. Three-dimensional bio-printing and bone tissue engineering: technical innovations and potential applications in maxillofacial reconstructive surgery. Maxillofac Plast Reconstr Surg. 2020;42(1):1–9.
3. Hynes K, Menicanin D, Gronthos S, Bartold PM. Clinical utility of stem cells for periodontal regeneration. Periodontol. 2012;59(1):203–27.
4. Robinton DA, Daley GQ. The promise of induced pluripotent stem cells in research and therapy. Nature. 2012;481(7381):295–305.
5. Tomar GB, Srivastava RK, Gupta N, Barhanpurkar AP, Pote ST, Jhaveri HM, Mishra GC, Wani MR. Human gingiva-derived mesenchymal stem cells are superior to bone marrow-derived mesenchymal stem cells for cell therapy in regenerative medicine. Biochem Biophys Res Commun. 2010;393(3):377–83.
6. Peng L, Jia Z, Yin X, Zhang X, Liu Y, Chen P, Ma K, Zhou C. Comparative analysis of mesenchymal stem cells from bone marrow, cartilage, and adipose tissue. Stem Cells Dev. 2008;17(4):761–74.
7. Thomas George V, Thomas NG, John S, Ittycheria PG. The scope of stem cells in periodontal regeneration. Tissue Eng. 2015;1:13–21.
8. Lee S-M, Zhang Q, Le AD. Dental stem cells: sources and potential applications. Curr Oral Health Rep. 2014;1(1):34–42.

9. Marynka-Kalmani K, Treves S, Yafee M, Rachima H, Gafni Y, Cohen MA, Pitaru S. The lamina propria of adult human oral mucosa harbors a novel stem cell population. Stem Cells. 2010;28(5):984–95.
10. Tang L, Li N, Xie H, Jin Y. Characterization of mesenchymal stem cells from human normal and hyperplastic gingiva. J Cell Physiol. 2011;226(3):832–42.
11. Seo B-M, Miura M, Gronthos S, Bartold PM, Batouli S, Brahim J, Young M, Robey PG, Wang CY, Shi S. Investigation of multipotent postnatal stem cells from human periodontal ligament. Lancet. 2004;364(9429):149–55.
12. Gay IC, Chen S, MacDougall M. Isolation and characterization of multipotent human periodontal ligament stem cells. Orthod Craniofac Res. 2007;10(3):149–60.
13. Maeda H, Wada N, Fujii S, Tomokiyo A, Akamine A. Periodontal ligament stem cells. In: Stem cells in clinic and research. IntechOpen; 2011.
14. Yao S, Pan F, Prpic V, Wise G. Differentiation of stem cells in the dental follicle. J Dent Res. 2008;87(8):767–71.
15. Honda MJ, Imaizumi M, Tsuchiya S, Morsczeck C. Dental follicle stem cells and tissue engineering. J Oral Sci. 2010;52(4):541–52.
16. Gronthos S, Brahim J, Li W, Fisher L, Cherman N, Boyde A, DenBesten P, Robey PG, Shi S. Stem cell properties of human dental pulp stem cells. J Dent Res. 2002;81(8):531–5.
17. Sloan AJ, Waddington RJ. Dental pulp stem cells: what, where, how? Int J Paediatr Dent. 2009;19(1):61–70.
18. Huang GT-J, Sonoyama W, Liu Y, Liu H, Wang S, Shi S. The hidden treasure in apical papilla: the potential role in pulp/dentin regeneration and bioroot engineering. J Endod. 2008;34(6):645–51.
19. Bakopoulou A, Leyhausen G, Volk J, Tsiftsoglou A, Garefis P, Koidis P, Geurtsen W. Comparative analysis of in vitro osteo/odontogenic differentiation potential of human dental pulp stem cells (DPSCs) and stem cells from the apical papilla (SCAP). Arch Oral Biol. 2011;56(7):709–21.
20. Sethi M, Dua A, Dodwad V. Stem cells: a window to regenerative dentistry. Int J Pharm Biomed Res. 2012;3(3):175–80.
21. Yang X, Ma Y, Guo W, Yang B, Tian W. Stem cells from human exfoliated deciduous teeth as an alternative cell source in bio-root regeneration. Theranostics. 2019;9(9):2694.
22. Iranparvar A, Nozariasbmarz A, DeGrave S, Tayebi L. Tissue engineering in periodontal regeneration. In: Applications of biomedical engineering in dentistry. Cham: Springer; 2020. p. 301–27.
23. Lynch S, Williams R, Poison A, Howell T, Reddy M, Zappa U, Antoniades H. A combination of platelet-derived and insulin-like growth factors enhances periodontal regeneration. J Clin Periodontol. 1989;16(8):545–8.
24. Rutherford R, Niekrash C, Kennedy J, Charette M. Platelet-derived and insulin-like growth factors stimulate regeneration of periodontal attachment in monkeys. J Periodontal Res. 1992;27(4):285–90.
25. Jeong Park Y, Moo Lee Y, Nae Park S, Yoon Sheen S, Pyoung Chung C, Lee SJ. Platelet derived growth factor releasing chitosan sponge for periodontal bone regeneration. Biomaterials. 2000;21(2):153–9.
26. Kaigler D, Avila G, Wisner-Lynch L, Nevins ML, Nevins M, Rasperini G, Lynch SE, Giannobile WV. Platelet-derived growth factor applications in periodontal and peri-implant bone regeneration. Expert Opin Biol Ther. 2011;11(3):375–85.
27. Sasikumar KP, Elavarasu S, Gadagi JS. The application of bone morphogenetic proteins to periodontal and peri-implant tissue regeneration: a literature review. J Pharm Bioallied Sci. 2012;4(Suppl 2):S427.
28. Anusuya GS, Kandasamy M, Raja SJ, Sabarinathan S, Ravishankar P, Kandhasamy B. Bone morphogenetic proteins: signaling periodontal bone regeneration and repair. J Pharm Bioallied Sci. 2016;8(Suppl 1):S39.
29. Kao RT, Murakami S, Beirne OR. The use of biologic mediators and tissue engineering in dentistry. Periodontol. 2009;50(1):127–53.

30. Murakami S. Periodontal tissue regeneration by signaling molecule (s): what role does basic fibroblast growth factor (FGF-2) have in periodontal therapy? Periodontol. 2011;56(1):188–208.

31. Lee A-R, Choi H, Kim J-H, Cho S-W, Park Y-B. Effect of serial use of bone morphogenetic protein 2 and fibroblast growth factor 2 on periodontal tissue regeneration. Implant Dent. 2017;26(5):664–73.

32. Janssens K, ten Dijke P, Janssens S, Van Hul W. Transforming growth factor-β1 to the bone. Endocr Rev. 2005;26(6):743–74.

33. Ogino Y, Ayukawa Y, Kukita T, Koyano K. The contribution of platelet-derived growth factor, transforming growth factor-β1, and insulin-like growth factor-I in platelet-rich plasma to the proliferation of osteoblast-like cells. Oral Surg Oral Med Oral Pathol Oral Radiol Endod. 2006;101(6):724–9.

34. Zhang Y, Cheng X, Wang J, Wang Y, Shi B, Huang C, Yang X, Liu T. Novel chitosan/collagen scaffold containing transforming growth factor-β1 DNA for periodontal tissue engineering. Biochem Biophys Res Commun. 2006;344(1):362–9.

35. Lee J-Y, Kim K-H, Shin S-Y, Rhyu I-C, Lee Y-M, Park Y-J, Chung C-P, Lee S-J. Enhanced bone formation by transforming growth factor-β1-releasing collagen/chitosan microgranules. J Biomed Mater Res A. 2006;76A(3):530–9.

36. Hammarström L. Enamel matrix, cementum development and regeneration. J Clin Periodontol. 1997;24(9):658–68.

37. Bosshardt DD. Biological mediators and periodontal regeneration: a review of enamel matrix proteins at the cellular and molecular levels. J Clin Periodontol. 2008;35:87–105.

38. Masaeli R, Zandsalimi K, Lotfi Z, Tayebi L. Using enamel matrix derivative to improve treatment efficacy in periodontal furcation defects. J Prosthodont. 2018;27(8):733–6.

39. Nishimura F, Terranova VP. Comparative study of the chemotactic responses of periodontal ligament cells and gingival fibroblasts to polypeptide growth factors. J Dent Res. 1996;75(4):986–92.

40. Dereka X, Markopoulou C, Vrotsos I. Role of growth factors on periodontal repair. Growth Factors. 2006;24(4):260–7.

41. Lee J, Stavropoulos A, Susin C, Wikesjö UM. Periodontal regeneration: focus on growth and differentiation factors. Dent Clin. 2010;54(1):93–111.

42. Okuda K, Kawase T, Momose M, Murata M, Saito Y, Suzuki H, Wolff LF, Yoshie H. Platelet-richplasma contains high levels of platelet-derived growth factor and transforming growthfactor-β and modulates the proliferation of periodontally related cells in vitro. J Periodontol. 2003;74(6):849–57.

43. Del Fabbro M, Bortolin M, Taschieri S, Weinstein R. Is platelet concentrate advantageous for the surgical treatment of periodontal diseases? A systematic review and meta-analysis. J Periodontol. 2011;82(8):1100–11.

44. Griffith LG, Naughton G. Tissue engineering--current challenges and expanding opportunities. Science. 2002;295(5557):1009–14.

45. Freed LE, Guilak F, Guo XE, Gray ML, Tranquillo R, Holmes JW, Radisic M, Sefton MV, Kaplan D, Vunjak-Novakovic G. Advanced tools for tissue engineering: scaffolds, bioreactors, and signaling. Tissue Eng. 2006;12(12):3285–305.

46. O'brien FJ. Biomaterials & scaffolds for tissue engineering. Mater Today. 2011;14(3):88–95.

47. Novosel EC, Kleinhans C, Kluger PJ. Vascularization is the key challenge in tissue engineering. Adv Drug Deliv Rev. 2011;63(4–5):300–11.

48. Roseti L, Parisi V, Petretta M, Cavallo C, Desando G, Bartolotti I, Grigolo B. Scaffolds for bone tissue engineering: state of the art and new perspectives. Mater Sci Eng C. 2017;78:1246–62.

49. Jazayeri HE, Tahriri M, Razavi M, Khoshroo K, Fahimipour F, Dashtimoghadam E, Almeida L, Tayebi L. A current overview of materials and strategies for potential use in maxillofacial tissue regeneration. Mater Sci Eng C. 2017;70:913–29.

50. Yazdimamaghani M, Razavi M, Vashaee D, Pothineni VR, Rajadas J, Tayebi L. Significant degradability enhancement in multilayer coating of polycaprolactone-bioactive glass/gelatin-

bioactive glass on magnesium scaffold for tissue engineering applications. Appl Surf Sci. 2015;338:137–45.
51. Razavi M, Fathi M, Savabi O, Razavi SM, Beni BH, Vashaee D, Tayebi L. Surface modification of magnesium alloy implants by nanostructured bredigite coating. Mater Lett. 2013;113:174–8.
52. Razavi M, Fathi MH, Savabi O, Vashaee D, Tayebi L. Biodegradation, bioactivity and in vivo biocompatibility analysis of plasma electrolytic oxidized (PEO) biodegradable mg implants. Phys Sci Int J. 2014;2014:708–22.
53. Yazdimamaghani M, Razavi M, Vashaee D, Tayebi L. Microstructural and mechanical study of PCL coated mg scaffolds. Surf Eng. 2014;30(12):920–6.
54. Alvarez K, Nakajima H. Metallic scaffolds for bone regeneration. Materials. 2009;2(3):790–832.
55. Yusop A, Bakir A, Shaharom N, Abdul Kadir M, Hermawan H. Porous biodegradable metals for hard tissue scaffolds: a review. Int J Biomater. 2012;2012:641430.
56. Yazdimamaghani M, Razavi M, Vashaee D, Moharamzadeh K, Boccaccini AR, Tayebi L. Porous magnesium-based scaffolds for tissue engineering. Mater Sci Eng C. 2017;71:1253–66.
57. Ramay HR, Zhang M. Biphasic calcium phosphate nanocomposite porous scaffolds for load-bearing bone tissue engineering. Biomaterials. 2004;25(21):5171–80.
58. Tripathi G, Basu B. A porous hydroxyapatite scaffold for bone tissue engineering: Physico-mechanical and biological evaluations. Ceram Int. 2012;38(1):341–9.
59. Salahinejad E, Hadianfard M, Macdonald D, Mozafari M, Vashaee D, Tayebi L. Zirconium titanate thin film prepared by an aqueous particulate sol–gel spin coating process using carboxymethyl cellulose as dispersant. Mater Lett. 2012;88:5–8.
60. Rouhani P, Salahinejad E, Kaul R, Vashaee D, Tayebi L. Nanostructured zirconium titanate fibers prepared by particulate sol–gel and cellulose templating techniques. J Alloys Compd. 2013;568:102–5.
61. Salahinejad E, Hadianfard M, Macdonald D, Karimi I, Vashaee D, Tayebi L. Aqueous sol–gel synthesis of zirconium titanate (ZrTiO4) nanoparticles using chloride precursors. Ceram Int. 2012;38(8):6145–9.
62. Mozafari M, Salahinejad E, Shabafrooz V, Yazdimamaghani M, Vashaee D, Tayebi L. Multilayer bioactive glass/zirconium titanate thin films in bone tissue engineering and regenerative dentistry. Int J Nanomedicine. 2013;8:1665.
63. Freed LE, Vunjak-Novakovic G, Biron RJ, Eagles DB, Lesnoy DC, Barlow SK, Langer R. Biodegradable polymer scaffolds for tissue engineering. Bio/Technology. 1994;12(7):689–93.
64. Ma PX, Choi J-W. Biodegradable polymer scaffolds with well-defined interconnected spherical pore network. Tissue Eng. 2001;7(1):23–33.
65. Place ES, George JH, Williams CK, Stevens MM. Synthetic polymer scaffolds for tissue engineering. Chem Soc Rev. 2009;38(4):1139–51.
66. Genova T, Roato I, Carossa M, Motta C, Cavagnetto D, Mussano F. Advances on bone substitutes through 3D bioprinting. Int J Mol Sci. 2020;21(19):7012.
67. Cancedda R, Giannoni P, Mastrogiacomo M. A tissue engineering approach to bone repair in large animal models and in clinical practice. Biomaterials. 2007;28(29):4240–50.
68. Nesic D, Schaefer BM, Sun Y, Saulacic N, Sailer I. 3D printing approach in dentistry: the future for personalized oral soft tissue regeneration. J Clin Med. 2020;9(7):2238.
69. Park S-H, Jung CS, Min B-H. Advances in three-dimensional bioprinting for hard tissue engineering. Tissue Eng Regen Med. 2016;13(6):622–35.
70. Sachs E, Cima M, Williams P, Brancazio D, Cornie J. Three dimensional printing: rapid tooling and prototypes directly from a CAD model. CIRP Ann. 1992;39:201–4.
71. Sanz AR, Carrión FS, Chaparro AP. Mesenchymal stem cells from the oral cavity and their potential value in tissue engineering. Periodontol. 2015;67(1):251–67.
72. Pagni G, Kaigler D, Rasperini G, Avila-Ortiz G, Bartel R, Giannobile W. Bone repair cells for craniofacial regeneration. Adv Drug Deliv Rev. 2012;64(12):1310–9.

73. Park J-Y, Jeon SH, Choung P-H. Efficacy of periodontal stem cell transplantation in the treatment of advanced periodontitis. Cell Transplant. 2011;20(2):271–86.
74. Chen F-M, Sun H-H, Lu H, Yu Q. Stem cell-delivery therapeutics for periodontal tissue regeneration. Biomaterials. 2012;33(27):6320–44.
75. Wang P, Berry D, Moran A, He F, Tam T, Chen L, Chen S. Controlled growth factor release in 3D-printed hydrogels. Adv Healthc Mater. 2020;9(15):1900977.
76. Zhu J, Marchant RE. Design properties of hydrogel tissue-engineering scaffolds. Expert Rev Med Devices. 2011;8(5):607–26.
77. Guvendiren M, Molde J, Soares RM, Kohn J. Designing biomaterials for 3D printing. ACS Biomater Sci Eng. 2016;2(10):1679–93.
78. Tamay DG, Dursun Usal T, Alagoz AS, Yucel D, Hasirci N, Hasirci V. 3D and 4D printing of polymers for tissue engineering applications. Front Bioeng Biotechnol. 2019;7:164.
79. Ma Y, Ji Y, Zhong T, Wan W, Yang Q, Li A, Zhang X, Lin M. Bioprinting-based PDLSC-ECM screening for in vivo repair of alveolar bone defect using cell-laden, injectable and photocrosslinkable hydrogels. ACS Biomater Sci Eng. 2017;3(12):3534–45.
80. You F, Chen X, Cooper D, Chang T, Eames BF. Homogeneous hydroxyapatite/alginate composite hydrogel promotes calcified cartilage matrix deposition with potential for three-dimensional bioprinting. Biofabrication. 2018;11(1):015015.
81. Chen M, Zhang Y, Xie Q, Zhang W, Pan X, Gu P, Zhou H, Gao Y, Walther A, Fan X. Long-term bone regeneration enabled by a polyhedral oligomeric silsesquioxane (POSS)-enhanced biodegradable hydrogel. ACS Biomater Sci Eng. 2019;5(9):4612–23.
82. Fahimipour F, Dashtimoghadam E, Hasani-Sadrabadi MM, Vargas J, Vashaee D, Lobner DC, Kashi TSJ, Ghasemzadeh B, Tayebi L. Enhancing cell seeding and osteogenesis of MSCs on 3D printed scaffolds through injectable BMP2 immobilized ECM-mimetic gel. Dent Mater. 2019;35(7):990–1006.
83. Cheng X, Yoo JJ, Hale RG, Davis MR, Kang H-W, Jin S. 3D printed biomaterials for maxillofacial tissue engineering and reconstruction—a review. Open J Biomed Mater Res. 2014;1:1–7.
84. Simmons CA, Alsberg E, Hsiong S, Kim WJ, Mooney DJ. Dual growth factor delivery and controlled scaffold degradation enhance in vivo bone formation by transplanted bone marrow stromal cells. Bone. 2004;35(2):562–9.
85. Cheng C-H, Chen Y-W, Lee AK-X, Yao C-H, Shie M-Y. Development of mussel-inspired 3D-printed poly (lactic acid) scaffold grafted with bone morphogenetic protein-2 for stimulating osteogenesis. J Mater Sci Mater Med. 2019;30(7):78.
86. Kumar PS, Hashimi S, Saifzadeh S, Ivanovski S, Vaquette C. Additively manufactured biphasic construct loaded with BMP-2 for vertical bone regeneration: a pilot study in rabbit. Mater Sci Eng C. 2018;92:554–64.
87. Golzar H, Mohammadrezaei D, Yadegari A, Rasoulianboroujeni M, Hashemi M, Omidi M, Yazdian F, Shalbaf M, Tayebi L. Incorporation of functionalized reduced graphene oxide/magnesium nanohybrid to enhance the osteoinductivity capability of 3D printed calcium phosphate-based scaffolds. Compos Part B. 2020;185:107749.
88. Rasoulianboroujeni M, Fahimipour F, Shah P, Khoshroo K, Tahriri M, Eslami H, Yadegari A, Dashtimoghadam E, Tayebi L. Development of 3D-printed PLGA/TiO2 nanocomposite scaffolds for bone tissue engineering applications. Mater Sci Eng C. 2019;96:105–13.
89. Fahimipour F, Dashtimoghadam E, Rasoulianboroujeni M, Yazdimamaghani M, Khoshroo K, Tahriri M, Yadegari A, Gonzalez JA, Vashaee D, Lobner DC. Collagenous matrix supported by a 3D-printed scaffold for osteogenic differentiation of dental pulp cells. Dent Mater. 2018;34(2):209–20.
90. Hindy A, Farahmand F, Pourdanesh F, Torshabi M, Al Janabi AH, Rasoulianboroujeni M, Tayebi L, Tabatabaei FS. Synthesis and characterization of 3D-printed functionally graded porous titanium alloy. J Mater Sci. 2020;55(21):9082–94.
91. Gopinathan J, Noh I. Recent trends in bioinks for 3D printing. Biomater Res. 2018;22(1):11.

92. Ribeiro A, Blokzijl MM, Levato R, Visser CW, Castilho M, Hennink WE, Vermonden T, Malda J. Assessing bioink shape fidelity to aid material development in 3D bioprinting. Biofabrication. 2017;10(1):014102.
93. Yan Y, Wang X, Xiong Z, Liu H, Liu F, Lin F, Wu R, Zhang R, Lu Q. Direct construction of a three-dimensional structure with cells and hydrogel. J Bioact Compat Polym. 2005;20(3):259–69.
94. Raveendran NT, Vaquette C, Meinert C, Ipe DS, Ivanovski S. Optimization of 3D bioprinting of periodontal ligament cells. Dent Mater. 2019;35(12):1683–94.
95. Lang NP, Lindhe J. Clinical periodontology and implant dentistry, 2 volume set. Wiley; 2015.
96. Thoma DS, Benić GI, Zwahlen M, Hämmerle CH, Jung RE. A systematic review assessing soft tissue augmentation techniques. Clin Oral Implants Res. 2009;20:146–65.
97. Benninger B, Andrews K, Carter W. Clinical measurements of hard palate and implications for subepithelial connective tissue grafts with suggestions for palatal nomenclature. J Oral Maxillofac Surg. 2012;70(1):149–53.
98. Del Pizzo M, Modica F, Bethaz N, Priotto P, Romagnoli R. The connective tissue graft: a comparative clinical evaluation of wound healing at the palatal donor site: a preliminary study. J Clin Periodontol. 2002;29(9):848–54.
99. Griffin TJ, Cheung WS, Zavras AI, Damoulis PD. Postoperative complications following gingival augmentation procedures. J Periodontol. 2006;77(12):2070–9.
100. Soileau KM, Brannon RB. A histologic evaluation of various stages of palatal healing following subepithelial connective tissue grafting procedures: a comparison of eight cases. J Periodontol. 2006;77(7):1267–73.
101. Zucchelli G, Mele M, Stefanini M, Mazzotti C, Marzadori M, Montebugnoli L, De Sanctis M. Patient morbidity and root coverage outcome after subepithelial connective tissue and de-epithelialized grafts: a comparative randomized-controlled clinical trial. J Clin Periodontol. 2010;37(8):728–38.
102. Aichelmann-Reidy ME, Yukna RA, Evans GH, Nasr HF, Mayer ET. Clinical evaluation of acellular allograft dermis for the treatment of human gingival recession. J Periodontol. 2001;72(8):998–1005.
103. Wei PC, Laurell L, Geivelis M, Lingen MW, Maddalozzo D. Acellular dermal matrix allografts to achieve increased attached gingiva. Part 1. A clinical study. J Periodontol. 2000;71(8):1297–305.
104. Wei PC, Laurell L, Lingen MW, Geivelis M. Acellular dermal matrix allografts to achieve increased attached gingiva. Part 2. A histological comparative study. J Periodontol. 2002;73(3):257–65.
105. Maiorana C, Beretta M, Pivetti L, Stoffella E, Grossi GB, Herford AS. Use of a collagen matrix as a substitute for free mucosal grafts in pre-prosthetic surgery: 1 year results from a clinical prospective study on 15 patients. Open Dent J. 2016;10:395.
106. Nevins M, Nevins ML, Kim S-W, Schupbach P, Kim DM. The use of mucograft collagen matrix to augment the zone of keratinized tissue around teeth: a pilot study. Int J Periodont Restor Dent. 2011;31(4):366–73.
107. Zuhr O, Bäumer D, Hürzeler M. The addition of soft tissue replacement grafts in plastic periodontal and implant surgery: critical elements in design and execution. J Clin Periodontol. 2014;41:S123–42.
108. Vignoletti F, Nunez J, Sanz M. Soft tissue wound healing at teeth, dental implants and the edentulous ridge when using barrier membranes, growth and differentiation factors and soft tissue substitutes. J Clin Periodontol. 2014;41:S23–35.
109. Allon I, Kaplan I, Gal G, Chaushu G, Allon DM. The clinical characteristics of benign oral mucosal tumors. Med Oral Patol Oral Cir Bucal. 2014;19(5):e438.
110. Chappuis V, Araújo MG, Buser D. Clinical relevance of dimensional bone and soft tissue alterations post-extraction in esthetic sites. Periodontol. 2017;73(1):73–83.
111. Pradeep K, Rajababu P, Satyanarayana D, Sagar V. Gingival recession: review and strategies in treatment of recession. Case Rep Dent. 2012;2012:563421.

112. Chambrone L, Tatakis DN. Long-term outcomes of untreated buccal gingival recessions: a systematic review and meta-analysis. J Periodontol. 2016;87(7):796–808.

113. Cairo F, Nieri M, Cincinelli S, Mervelt J, Pagliaro U. The interproximal clinical attachment level to classify gingival recessions and predict root coverage outcomes: an explorative and reliability study. J Clin Periodontol. 2011;38(7):661–6.

114. Muthukumar S, Rangarao S. Surgical augmentation of interdental papilla-a case series. Contemp Clin Dent. 2015;6(Suppl 1):S294.

115. Prato GPP, Rotundo R, Cortellini P, Tinti C, Azzi R. Interdental papilla management: a review and classification of the therapeutic approaches. J Prosthet Dent. 2004;92(5):476.

116. Ahila E, Kumar RS, Reddy VK, Pratebha B, Jananni M, Priyadharshini V. Augmentation of interdental papilla with platelet-rich fibrin. Contemp Clin Dent. 2018;9(2):213.

117. Awartani FA, Tatakis DN. Interdental papilla loss: treatment by hyaluronic acid gel injection: a case series. Clin Oral Investig. 2016;20(7):1775–80.

118. Carranza N, Zogbi C. Reconstruction of the interdental papilla with an underlying subepithelial connective tissue graft: technical considerations and case reports. Int J Periodontics Restorative Dent. 2011;31(5):533.

119. Feuillet D, Keller J-F, Agossa K. Interproximal tunneling with a customized connective tissue graft: a microsurgical technique for interdental papilla reconstruction. Int J Periodont Restor Dent. 2018;38(6):833–9.

120. Akiyama K. Papilla reconstruction using the dental operating microscope. MICRO Int J Microdent. 2009;1:25–9.

121. Lee WP, Kim HJ, Yu SJ, Kim BO. Six month clinical evaluation of interdental papilla reconstruction with injectable hyaluronic acid gel using an image analysis system. J Esthet Restor Dent. 2016;28(4):221–30.

122. Spano SJ, Ghilzon R, Lam DK, Goldberg MB, Tenenbaum HC. Subperiosteal papilla augmentation with a non–animal-derived hyaluronic acid overlay technique. Clin Adv Periodont. 2020;10(1):4–9.

123. Carnio J. Surgical reconstruction of interdental papilla using an interposed subepithelial connective tissue graft: a case report. J Prosthet Dent. 2004;92(3):282.

124. Tayebi L, Rasoulianboroujeni M, Cui Z, Ye H. 3D-printed thick structured gelatin membrane for engineering of heterogeneous tissues. Mater Lett. 2018;217:39–43.

125. Toledo Stuani VD, do Prado Manfredi GG, Kondo VAM, Noritomi PY, Lisboa-Filho PN, Sant'Ana ACP. The use of additively manufactured scaffolds for treating gingival recession associated with interproximal defects. J 3D Print Med. 2020;4(3):153–65.

126. Nesic D, Durual S, Marger L, Mekki M, Sailer I, Scherrer SS. Could 3D printing be the future for oral soft tissue regeneration? Bioprinting. 2020;2020:e00100.

127. Schemitsch EH. Size matters: defining critical in bone defect size! J Orthop Trauma. 2017;31:S20–2.

128. Gómez-Barrena E, Rosset P, Lozano D, Stanovici J, Ermthaller C, Gerbhard F. Bone fracture healing: cell therapy in delayed unions and nonunions. Bone. 2015;70:93–101.

129. Masaeli R, Zandsalimi K, Rasoulianboroujeni M, Tayebi L. Challenges in three-dimensional printing of bone substitutes. Tissue Eng Part B Rev. 2019;25(5):387–97.

130. Tayebi L, Rasoulianboroujeni M. https://patents.google.com/patent/WO2017192525A1/en, WO2017192525A1, PCT/US2017/030555, Reinforced Bone Scaffold, US, International. 2017.

131. Nejati E, Mirzadeh H, Zandi M. Synthesis and characterization of nano-hydroxyapatite rods/poly (l-lactide acid) composite scaffolds for bone tissue engineering. Compos A: Appl Sci Manuf. 2008;39(10):1589–96.

132. Lee SJ, Lim GJ, Lee J-W, Atala A, Yoo JJ. In vitro evaluation of a poly (lactide-co-glycolide)–collagen composite scaffold for bone regeneration. Biomaterials. 2006;27(18):3466–72.

133. Fei Z, Hu Y, Wu D, Wu H, Lu R, Bai J, Song H. Preparation and property of a novel bone graft composite consisting of rhBMP-2 loaded PLGA microspheres and calcium phosphate cement. J Mater Sci Mater Med. 2008;19(3):1109–16.

134. Zhao J, Guo L, Yang X, Weng J. Preparation of bioactive porous HA/PCL composite scaffolds. Appl Surf Sci. 2008;255(5):2942–6.
135. Thein-Han W, Misra R. Biomimetic chitosan–nanohydroxyapatite composite scaffolds for bone tissue engineering. Acta Biomater. 2009;5(4):1182–97.
136. Mozafari M, Rabiee M, Azami M, Maleknia S. Biomimetic formation of apatite on the surface of porous gelatin/bioactive glass nanocomposite scaffolds. Appl Surf Sci. 2010;257(5):1740–9.
137. Sabir MI, Xu X, Li L. A review on biodegradable polymeric materials for bone tissue engineering applications. J Mater Sci. 2009;44(21):5713–24.
138. Ikeda R, Fujioka H, Nagura I, Kokubu T, Toyokawa N, Inui A, Makino T, Kaneko H, Doita M, Kurosaka M. The effect of porosity and mechanical property of a synthetic polymer scaffold on repair of osteochondral defects. Int Orthop. 2009;33(3):821–8.
139. Atala A, Lanza R, Mikos T, Nerem R. Principles of regenerative medicine. Academic Press; 2018.
140. Roberts A, Brackley C. Friction of surgeons' gloves. J Phys D Appl Phys. 1992;25(1A):A28.
141. Keogh MB, O'Brien FJ, Daly JS. A novel collagen scaffold supports human osteogenesis—applications for bone tissue engineering. Cell Tissue Res. 2010;340(1):169–77.
142. Al-Munajjed AA, Plunkett NA, Gleeson JP, Weber T, Jungreuthmayer C, Levingstone T, Hammer J, O'Brien FJ. Development of a biomimetic collagen-hydroxyapatite scaffold for bone tissue engineering using a SBF immersion technique. J Biomed Mater Res B Appl Biomater. 2009;90B(2):584–91.
143. Hoyer B, Bernhardt A, Heinemann S, Stachel I, Meyer M, Gelinsky M. Biomimetically mineralized salmon collagen scaffolds for application in bone tissue engineering. Biomacromolecules. 2012;13(4):1059–66.
144. Tayebi L, To be published 2021.
145. Turner CH, Burr DB. Basic biomechanical measurements of bone: a tutorial. Bone. 1993;14(4):595–608.
146. Parthasarathy J. 3D modeling, custom implants and its future perspectives in craniofacial surgery. Ann Maxillofac Surg. 2014;4(1):9.
147. Ying T, Wang DM, Tong J, Wang CT, Zhang CP. Three-dimensional finite-element analysis investigating the biomechanical effects of human mandibular reconstruction with autogenous bone grafts. J Cranio-Maxillofac Surg. 2006;34(5):290–8.
148. Mahdian N, Dostálová TJ, Daněk J, Nedoma J, Kohout J, Hubáček M, Hliňáková P. 3D reconstruction of TMJ after resection of the cyst and the stress–strain analyses. Comput Methods Prog Biomed. 2013;110(3):279–89.
149. Sobotta J. Atlas de anatomía humana: Cabeza, cuello, miembro superior, vol. 1. Médica Panamericana; 2006.
150. Gregolin RF, Zavaglia CADC, Tokimatsu RC, Pereira JA. Biomechanical stress and strain analysis of mandibular human region from computed tomography to custom implant development. Adv Mater Sci Eng. 2017;2017:1–9.
151. Omidi M, Almeida LE, Tayebi L. To be published. 2021.
152. Cassell CO, Hofer OS, Morrison WA, Knight KR. Vascularisation of tissue-engineered grafts: the regulation of angiogenesis in reconstructive surgery and in disease states. Br J Plast Surg. 2002;55(8):603–10.
153. Yu H, VandeVord PJ, Mao L, Matthew HW, Wooley PH, Yang S-Y. Improved tissue-engineered bone regeneration by endothelial cell mediated vascularization. Biomaterials. 2009;30(4):508–17.
154. Schmidmaier G, Capanna R, Wildemann B, Beque T, Lowenberg D. Bone morphogenetic proteins in critical-size bone defects: what are the options? Injury. 2009;40:S39–43.
155. Dimitriou R, Jones E, McGonagle D, Giannoudis PV. Bone regeneration: current concepts and future directions. BMC Med. 2011;9(1):66.
156. Khan WS, Rayan F, Dhinsa BS, Marsh D. An osteoconductive, osteoinductive, and osteogenic tissue-engineered product for trauma and orthopaedic surgery: how far are we? Stem Cells Int. 2012;2012:236231.

157. Hernigou P. Bone transplantation and tissue engineering, part III: allografts, bone grafting and bone banking in the twentieth century. Int Orthop. 2015, 2015:1–11.
158. Aponte-Tinao LA, Ayerza MA, Muscolo DL, Farfalli GL. What are the risk factors and management options for infection after reconstruction with massive bone allografts? Clin Orthopaed Relat Res. 2015;2015:1–5.
159. Vormoor B, Knizia HK, Batey MA, Almeida GS, Wilson I, Dildey P, Sharma A, Blair H, Hide IG, Heidenreich O. Development of a preclinical orthotopic xenograft model of Ewing sarcoma and other human malignant bone disease using advanced in vivo imaging. PLoS One. 2014;9(1):e85128.
160. Ardakani AG, Cheema U, Brown RA, Shipley RJ. Quantifying the correlation between spatially defined oxygen gradients and cell fate in an engineered three-dimensional culture model. J R Soc Interface. 2014;11(98):20140501.
161. Lenze U, Pohlig F, Seitz S, Ern C, Milz S, Docheva D, Schieker M. Influence of osteogenic stimulation and VEGF treatment on in vivo bone formation in hMSC-seeded cancellous bone scaffolds. BMC Musculoskelet Disord. 2014;15(1):350.
162. Volkmer E, Drosse I, Otto S, Stangelmayer A, Stengele M, Kallukalam BC, Mutschler W, Schieker M. Hypoxia in static and dynamic 3D culture systems for tissue engineering of bone. Tissue Eng A. 2008;14(8):1331–40.
163. Santos MI, Reis RL. Vascularization in bone tissue engineering: physiology, current strategies, major hurdles and future challenges. Macromol Biosci. 2010;10(1):12–27.
164. Nguyen LH, Annabi N, Nikkhah M, Bae H, Binan L, Park S, Kang Y, Yang Y, Khademhosseini A. Vascularized bone tissue engineering: approaches for potential improvement. Tissue Eng Part B Rev. 2012;18(5):363–82.
165. Zisch AH, Lutolf MP, Ehrbar M, Raeber GP, Rizzi SC, Davies N, Schmökel H, Bezuidenhout D, Djonov V, Zilla P. Cell-demanded release of VEGF from synthetic, biointeractive cell ingrowth matrices for vascularized tissue growth. FASEB J. 2003;17(15):2260–2.
166. Ishijima M, Suzuki N, Hozumi K, Matsunobu T, Kosaki K, Kaneko H, Hassell JR, Arikawa-Hirasawa E, Yamada Y. Perlecan modulates VEGF signaling and is essential for vascularization in endochondral bone formation. Matrix Biol. 2012;31(4):234–45.
167. Richardson TP, Peters MC, Ennett AB, Mooney DJ. Polymeric system for dual growth factor delivery. Nat Biotechnol. 2001;19(11):1029–34.
168. Patel ZS, Young S, Tabata Y, Jansen JA, Wong ME, Mikos AG. Dual delivery of an angiogenic and an osteogenic growth factor for bone regeneration in a critical size defect model. Bone. 2008;43(5):931–40.
169. Young S, Patel ZS, Kretlow JD, Murphy MB, Mountziaris PM, Baggett LS, Ueda H, Tabata Y, Jansen JA, Wong M. Dose effect of dual delivery of vascular endothelial growth factor and bone morphogenetic protein-2 on bone regeneration in a rat critical-size defect model. Tissue Eng A. 2009;15(9):2347–62.
170. Shah NJ, Macdonald ML, Beben YM, Padera RF, Samuel RE, Hammond PT. Tunable dual growth factor delivery from polyelectrolyte multilayer films. Biomaterials. 2011;32(26):6183–93.
171. Arkudas A, Tjiawi J, Bleiziffer O, Grabinger L, Polykandriotis E, Beier JP, Stürzl M, Horch RE, Kneser U. Fibrin gel-immobilized VEGF and bFGF efficiently stimulate angiogenesis in the AV loop model. Mol Med. 2007;13(9–10):480.
172. Elia R, Fuegy PW, VanDelden A, Firpo MA, Prestwich GD, Peattie RA. Stimulation of in vivo angiogenesis by in situ crosslinked, dual growth factor-loaded, glycosaminoglycan hydrogels. Biomaterials. 2010;31(17):4630–8.
173. Hu K, Olsen BR. The roles of vascular endothelial growth factor in bone repair and regeneration. Bone. 2016;91:30–8.
174. Hu K, Olsen BR. Osteoblast-derived VEGF regulates osteoblast differentiation and bone formation during bone repair. J Clin Invest. 2016;126(2):509–26.
175. Wang Y, Wan C, Deng L, Liu X, Cao X, Gilbert SR, Bouxsein ML, Faugere M-C, Guldberg RE, Gerstenfeld LC. The hypoxia-inducible factor α pathway couples angiogenesis to osteogenesis during skeletal development. J Clin Invest. 2007;117(6):1616–26.

176. Duan X, Bradbury SR, Olsen BR, Berendsen AD. VEGF stimulates intramembranous bone formation during craniofacial skeletal development. Matrix Biol. 2016;52:127–40.
177. Hill C, Jacobs B, Kennedy L, Rohde S, Zhou B, Baldwin S, Goudy S. Cranial neural crest deletion of VEGFa causes cleft palate with aberrant vascular and bone development. Cell Tissue Res. 2015;361(3):711–22.
178. Liu Y, Olsen BR. Distinct VEGF functions during bone development and homeostasis. Arch Immunol Ther Exp. 2014;62(5):363–8.
179. Liu Y, Berendsen AD, Jia S, Lotinun S, Baron R, Ferrara N, Olsen BR. Intracellular VEGF regulates the balance between osteoblast and adipocyte differentiation. J Clin Invest. 2012;122(9):3101–13.
180. Berendsen A, Olsen B. Regulation of adipogenesis and osteogenesis in mesenchymal stem cells by vascular endothelial growth factor a. J Intern Med. 2015;277(6):674–80.
181. Ramasamy SK, Kusumbe AP, Wang L, Adams RH. Endothelial notch activity promotes angiogenesis and osteogenesis in bone. Nature. 2014;507(7492):376–80.
182. Helmrich U, Di Maggio N, Güven S, Groppa E, Melly L, Largo RD, Heberer M, Martin I, Scherberich A, Banfi A. Osteogenic graft vascularization and bone resorption by VEGF-expressing human mesenchymal progenitors. Biomaterials. 2013;34(21):5025–35.
183. Tomanek RJ, Lotun K, Clark EB, Suvarna PR, Hu N. VEGF and bFGF stimulate myocardial vascularization in embryonic chick. Am J Phys Heart Circ Phys. 1998;274(5):H1620–6.
184. Ozawa CR, Banfi A, Glazer NL, Thurston G, Springer ML, Kraft PE, McDonald DM, Blau HM. Microenvironmental VEGF concentration, not total dose, determines a threshold between normal and aberrant angiogenesis. J Clin Invest. 2004;113(4):516–27.
185. Sheridan M, Shea L, Peters M, Mooney D. Bioabsorbable polymer scaffolds for tissue engineering capable of sustained growth factor delivery. J Control Release. 2000;64(1):91–102.
186. Chen RR, Mooney DJ. Polymeric growth factor delivery strategies for tissue engineering. Pharm Res. 2003;20(8):1103–12.
187. Lee J-Y, Nam S-H, Im S-Y, Park Y-J, Lee Y-M, Seol Y-J, Chung C-P, Lee S-J. Enhanced bone formation by controlled growth factor delivery from chitosan-based biomaterials. J Control Release. 2002;78(1):187–97.
188. Lee K, Silva EA, Mooney DJ. Growth factor delivery-based tissue engineering: general approaches and a review of recent developments. J R Soc Interface. 2011;8(55):153–70.
189. Holland TA, Tabata Y, Mikos AG. Dual growth factor delivery from degradable oligo (poly (ethylene glycol) fumarate) hydrogel scaffolds for cartilage tissue engineering. J Control Release. 2005;101(1):111–25.
190. Babensee JE, McIntire LV, Mikos AG. Growth factor delivery for tissue engineering. Pharm Res. 2000;17(5):497–504.
191. Yu X, Khalil A, Dang PN, Alsberg E, Murphy WL. Multilayered inorganic microparticles for tunable dual growth factor delivery. Adv Funct Mater. 2014;24(20):3082–93.
192. Kolambkar YM, Dupont KM, Boerckel JD, Huebsch N, Mooney DJ, Hutmacher DW, Guldberg RE. An alginate-based hybrid system for growth factor delivery in the functional repair of large bone defects. Biomaterials. 2011;32(1):65–74.
193. Thomopoulos S, Sakiyama-Elbert S, Silva M, Gelberman R, Xia Y, Schwartz A, Xie J, Polymer nanofiber scaffold for a heparin/fibrin based growth factor delivery system. Google Patents. 2014.
194. Lu S, Lam J, Trachtenberg JE, Lee EJ, Seyednejad H, van den Beucken JJ, Tabata Y, Wong ME, Jansen JA, Mikos AG. Dual growth factor delivery from bilayered, biodegradable hydrogel composites for spatially-guided osteochondral tissue repair. Biomaterials. 2014;35(31):8829–39.
195. Sandsdrap P, Moës A-J. Influence of manufacturing parameters on the size characteristics and the release profiles of nifedipine from poly (DL-lactide-co-glycolide) microspheres. Int J Pharm. 1993;98(1):157–64.
196. Khojasteh A, Fahimipour F, Eslaminejad MB, Jafarian M, Jahangir S, Bastami F, Tahriri M, Karkhaneh A, Tayebi L. Development of PLGA-coated β-TCP scaffolds containing VEGF for bone tissue engineering. Mater Sci Eng C. 2016;69:780–8.

197. Davies N, Dobner S, Bezuidenhout D, Schmidt C, Beck M, Zisch AH, Zilla P. The dosage dependence of VEGF stimulation on scaffold neovascularisation. Biomaterials. 2008;29(26):3531–8.
198. Galvani L. De viribus electricitatis in motu musculari comentarius cum joannis aldini dissertatione et notis; accesserunt epistolae ad animalis electricitatis theoriam pertinentes. Apud Societatem Typographicam; 1792.
199. Fukada E, Yasuda I. On the piezoelectric effect of bone. J Phys Soc Jpn. 1957;12(10):1158–62.
200. Bassett CAL, Pawluk RJ, BECKER RO. Effects of electric currents on bone in vivo. Nature. 1964;204(4959):652–4.
201. Lavine LS, Lustrin I, Rinaldi RA, Shamos MH, Liboff AR. Electric enhancement of bone healing. Science. 1972;175(4026):1118–21.
202. Paterson D, Carter R, Maxwell G, Hillier T, Ludbrook J, Savage J. Electrical bone-growth stimulation in an experimental model of delayed union. Lancet. 1977;309(8025):1278–81.
203. Aubin JE, Gupta AK, Bhargava U, Turksen K. Expression and regulation of galectin 3 in rat osteoblastic cells. J Cell Physiol. 1996;169(3):468–80.
204. Malaval L, Modrowski D, Gupta AK, Aubin JE. Cellular expression of bone-related proteins during in vitro osteogenesis in rat bone marrow stromal cell cultures. J Cell Physiol. 1994;158(3):555–72.
205. Pinero GJ, Farach-Carson MC, Devoll RE, Aubin JE, Brunn JC, Butler WT. Bone matrix proteins in osteogenesis and remodelling in the neonatal rat mandible as studied by immunolocalization of osteopontin, bone sialoprotein, α2HS-glycoprotein and alkaline phosphatase. Arch Oral Biol. 1995;40(2):145–55.
206. Rodan G, Noda M. Gene expression in osteoblastic cells. Crit Rev Eukaryot Gene Expr. 1991;1(2):85–98.
207. Bendrea A-D, Cianga L, Cianga I. Progress in the field of conducting polymers for tissue engineering applications. J Biomater Appl. 2011;26(1):3–84.
208. Mozafari M, Mehraien M, Vashaee D, Tayebi L. Electroconductive nanocomposite scaffolds: a new strategy into tissue engineering and regenerative medicine. INTECH Open Access Publisher; 2012. p. 369–92.
209. Huang Z-B, Yin G-F, Liao X-M, Gu J-W. Conducting polypyrrole in tissue engineering applications. Front Mater Sci. 2014;8(1):39–45.
210. Gelmi A, Higgins MJ, Wallace GG. Physical surface and electromechanical properties of doped polypyrrole biomaterials. Biomaterials. 2010;31(8):1974–83.
211. Ateh D, Navsaria H, Vadgama P. Polypyrrole-based conducting polymers and interactions with biological tissues. J R Soc Interface. 2006;3(11):741–52.
212. Schmidt CE, Shastri VR, Vacanti JP, Langer R. Stimulation of neurite outgrowth using an electrically conducting polymer. Proc Natl Acad Sci. 1997;94(17):8948–53.
213. Williams R, Doherty P. A preliminary assessment of poly (pyrrole) in nerve guide studies. J Mater Sci Mater Med. 1994;5(6–7):429–33.
214. Cui X, Wiler J, Dzaman M, Altschuler RA, Martin DC. In vivo studies of polypyrrole/peptide coated neural probes. Biomaterials. 2003;24(5):777–87.
215. Jiang X, Marois Y, Traoré A, Tessier D, Dao LH, Guidoin R, Zhang Z. Tissue reaction to polypyrrole-coated polyester fabrics: an in vivo study in rats. Tissue Eng. 2002;8(4):635–47.
216. Wang X, Gu X, Yuan C, Chen S, Zhang P, Zhang T, Yao J, Chen F, Chen G. Evaluation of biocompatibility of polypyrrole in vitro and in vivo. J Biomed Mater Res Part A. 2004;68(3):411–22.
217. Ramanaviciene A, Kausaite A, Tautkus S, Ramanavicius A. Biocompatibility of polypyrrole particles: an in-vivo study in mice. J Pharm Pharmacol. 2007;59(2):311–5.
218. Babaie A, Bakhshandeh B, Abedi A, Mohammadnejad J, Shabani I, Ardeshirylajimi A, Moosavi SR, Amini J, Tayebi L. Synergistic effects of conductive PVA/PEDOT electrospun scaffolds and electrical stimulation for more effective neural tissue engineering. Eur Polym J. 2020;140:110051.

219. Shahini A, Yazdimamaghani M, Walker KJ, Eastman MA, Hatami-Marbini H, Smith BJ, Ricci JL, Madihally SV, Vashaee D, Tayebi L. 3D conductive nanocomposite scaffold for bone tissue engineering. Int J Nanomedicine. 2014;9:167.
220. Abedi A, Hasanzadeh M, Tayebi L. Conductive nanofibrous chitosan/PEDOT: PSS tissue engineering scaffolds. Mater Chem Phys. 2019;237:121882.
221. Yazdimamaghani M, Razavi M, Mozafari M, Vashaee D, Kotturi H, Tayebi L. Biomineralization and biocompatibility studies of bone conductive scaffolds containing poly (3, 4-ethylenedioxythiophene): poly (4-styrene sulfonate)(PEDOT: PSS). J Mater Sci Mater Med. 2015;26(12):1–11.
222. Tahmasbi Rad A, Ali N, Kotturi HSR, Yazdimamaghani M, Smay J, Vashaee D, Tayebi L. Conducting scaffolds for liver tissue engineering. J Biomed Mater Res A. 2014;102(11):4169–81.
223. Fani N, Hajinasrollah M, Asghari Vostikolaee M, Baghaban Eslaminejad M, Mashhadiabbas F, Tongas N, Rasoulianboroujeni M, Yadegari A, Ede K, Tahriri M. Influence of conductive PEDOT: PSS in a hard tissue scaffold: in vitro and in vivo study. J Bioact Compat Polym. 2019;34(6):436–41.
224. Yagci Y, Toppare L. Electroactive macromonomers based on pyrrole and thiophene: a versatile route to conducting block and graft polymers. Polym Int. 2003;52(10):1573–8.
225. McCullough LA, Matyjaszewski K. Conjugated conducting polymers as components in block copolymer systems. Mol Cryst Liq Cryst. 2010;521(1):1–55.
226. Guo B, Ma PX. Conducting polymers for tissue engineering. Biomacromolecules. 2018;19(6):1764–82.
227. Qazi TH, Rai R, Boccaccini AR. Tissue engineering of electrically responsive tissues using polyaniline based polymers: a review. Biomaterials. 2014;35(33):9068–86.
228. Mattioli-Belmonte M, Giavaresi G, Biagini G, Virgili L, Giacomini M, Fini M, Giantomassi F, Natali D, Torricelli P, Giardino R. Tailoring biomaterial compatibility: in vivo tissue response versus in vitro cell behavior. Int J Artif Organs. 2003;26(12):1077–85.
229. Shupak NM, Prato FS, Thomas AW. Therapeutic uses of pulsed magnetic-field exposure: a review. URSI Radio Sci Bull. 2003;2003(307):9–32.
230. Steenhuis H-J, Fang X, Ulusemre T. Global diffusion of innovation during the fourth industrial revolution: the case of additive manufacturing or 3D printing. Int J Innov Technol Manag. 2020;17(01):2050005.
231. Kirillova A, Bushev S, Abubakirov A, Sukikh G. Bioethical and legal issues in 3D bioprinting. Int J Bioprint. 2020;6(3):272.
232. Vermeulen N, Haddow G, Seymour T, Faulkner-Jones A, Shu W. 3D bioprint me: a socioethical view of bioprinting human organs and tissues. J Med Ethics. 2017;43(9):618–24.
233. Vijayavenkataraman S, Lu W, Fuh J. 3D bioprinting–an ethical, legal and social aspects (ELSA) framework. Bioprinting. 2016;1:11–21.

Application of 3D Printing in Reconstruction of Oral and Maxillofacial Multi- and Interfacial Tissue Defects

7

7.1 Introduction

Different tissue engineering approaches have been successfully used to fabricate three-dimensional tissue-resembling structures. However, the fabrication of multi- and interfacial tissue structures is still challenging.

Some important multi- and interfacial tissues in oral and maxillofacial surgeries include osteomucosal, osteochondral, and periodontal tissues as well as dentin/pulp complexes in a whole tooth. The regeneration of these constructs is inherently challenging because several cells, signaling molecules, and scaffolds are required to establish the natural hierarchical structures and interfacial connections. A major challenge in the regeneration of multi-tissue and interfacial constructs is integration of the soft/hard tissues as a strong interface is essential for the stability of the construct.

The success of regenerating functional tissue hinges on the ability to replicate the interfacial regions that connect hard and soft tissue [1]. Although significant progress has been made in the field of multi-tissue engineering, there is still a lack of focus on the soft/hard interfacial regions [2, 3]. Until sufficient regeneration of a functional interface, due to lack of suitable tissue adhesion method, is accomplished, the clinical applications of regenerated tissue will be severely limited [4].

The current approach for the integration of different tissues in multi-tissue constructs is the use of biocompatible natural adhesives such as fibrin glue and collagen adhesive [5]. However, their adhesion strength will diminish over time in a physiological condition [5, 6]. Moreover, they may act as a barrier and block direct cellular communications between tissues [5, 6]. More specifically, the addition of the adhesive as a third layer/material between the tissues disrupts the interfacial cellular communications [7, 8].

The other approach is using physical adhesion such as heat or solvent treatment, but this method restricts the capability for cell growth [2]. Suturing, another approach of adhesion, is often not possible in tissue engineering since most of the soft tissues are not sutureable [9, 10].

© Springer Nature Switzerland AG 2021
L. Tayebi et al., *3D Printing in Oral & Maxillofacial Surgery*,
https://doi.org/10.1007/978-3-030-77787-6_7

In this chapter, we discuss a biocompatible chemical process as a methodology to directly integrate soft and hard scaffolds.

The demands and potential applications of multi-tissue constructs in oral and maxillofacial surgery include:

1. Clinical applications: Tissue loss in oral and maxillofacial area is a commonly encountered problem that can range from small periodontal defects to more complex, difficult to manage structural defects. Data from National Health and Nutrition Examination Surveys showed a high incidence of periodontitis in the United States affecting approximately half of the population aged ≥30 years with 8.9% having severe periodontitis that warrants surgery [11]. A recent multicenter trauma research study, which was carried out across several European Oral Surgery centers, showed an increasing incidence of maxillofacial trauma resulting in bone loss [12]. It is estimated that 24,000 patients are diagnosed with oral cavity cancer in the United States each year [13], and oral cavity cancer ranks among the top ten most prevalent malignancies affecting patients worldwide [14]. In most of these cases, the surgeons need to do partial-thickness mandibular resection or marginal mandibulectomy [15]. In almost all of the above conditions, the surgeons have to deal with multi-tissue loss. In brief, multi-tissue construct can be used as a suitable graft material for intraoral and extraoral repair and treatment of soft and hard tissue defects, resulting from conditions such as cleft palate, trauma, and oral cancer. Intraoral applications include regeneration of periodontal defects, ridge augmentation prior to dental implant placement in atrophic bone and management of defects following trauma, surgical resection of cancer, and repair of cleft palate [16, 17]. Soft/hard tissue constructs are particularly useful in situations where there is a shortage of soft tissue to cover the bone graft in order to achieve primary wound closure. Some potential applications are similar to those of the barrier membranes used for Guided Tissue Engineering (GTR).

   Multi-tissue construct can also have potential for extraoral applications, including orbital floor reconstruction; spinal fusion surgery; orthopedic applications for regeneration of radial bone, rib, and femoral bone defects; and neurosurgical applications such as prevention of cerebrospinal fluid leakage from dura mater sites [18].

2. In vitro model to replace animal model: Multi-tissue constructs can be used as a relevant in vitro 3D model to replace animal models in order to assess the biological interaction of the oral tissues with various biomaterials and to evaluate dental implants and oral care products.

3. Oral disease modeling and drug delivery systems: Multi-tissue constructs can be used in the investigation of oral tissue disease phenomena, their progression, and treatment, including evaluation of drug delivery systems.

This chapter begins with discussing the role of 3D printing in regeneration of heterogenous interface. Next section is dedicated to osteomucosal tissue regeneration. In this section, after explaining the general progress in the field, a 3D printed osteomucosal design will be presented. Moreover, a mechanism of adhesion between hard and soft tissues scaffolds will be described that can be used for different multi-tissue constructs. Osteochondral tissue regeneration in the oral and maxillofacial region (e.g., mandibular condyle), periodontal tissue, and the different methods in its regeneration as well as the whole tooth regeneration will also be discussed in this chapter. Guided bone/tissue regeneration (GBR/GTR) 3D printed membranes and their role in oral and maxillofacial surgery will be the subject of the last section.

7.2 Regeneration of Heterogenous Interface Using 3D Printing Approach

In 2018, our group had suggested the use of 3D printing method to fabricate membranes that can be used for regeneration of heterogenous interface tissues and heterogeneous membrane-shape organs [19]. The concept was relying on the below facts:

1. 3D printing is a technique established on layer-by-layer deposition of materials. Thus, in contrast to traditional methods, such as molding, making structures with different configurations on different sides is easily achievable by 3D printing.
2. The two sides of a heterogenous interface are composed of two different tissues. These two tissues often need different environment to be repaired and regenerate properly. When we are using a membrane or thin scaffold to regenerate such tissues, it is highly effective to have different layers with different configurations (such as porosity, pore size, and materials) in two different sides to house the new cells grown at the interface area for joining the adjacent tissues.

The two above facts imply that 3D printing can offer a promising approach to accommodate the specific needs for regeneration of heterogenous interface.

Figure 7.1 illustrated the macroscopic image of such membranes made of gelatin. These 3D printed membranes were fabricated with different pore size and porosity on different sides. The authors suggested that such membranes may have potentials to be used in regeneration of full-thickness oral mucosa [19]. For this purpose, the thickness of the membrane was designed to be in the range of 0.2–2 mm. In the specific shape of Fig. 7.1c, d, the thickness was 300 μm made of six layers with the distance between strands of approximately 300 μm, 300 μm, 450 μm, 450 μm, 600 μm, and 600 μm in layers 1 to 6, respectively. The distance between the strands was decreased after cross-linking and drying, however, the gradient pore size configuration was remained. More specifically, the first later had the pore size of 100 μm approximately and the last layers comprised the pore size of 400 μm approximately. The angels between the strands were 90° in Fig. 7.1c, d. However, if smaller pore size was required, the angels of the strand-printing could be changed from 0° and 90° to 45° and 135°. Generally, altering the angles of strands in 3D printing procedures and using their different combinations can noticeably decrease the pore size. For example, in Fig. 7.1e, the last two layers were printed with 45°/135° angles to make a modified configuration with significantly smaller pore size on one side [19].

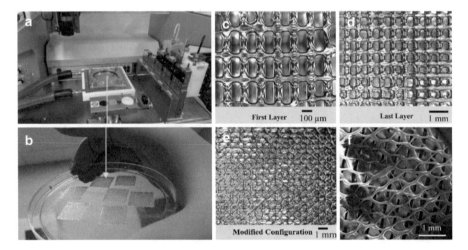

Fig. 7.1. The set-up of the 3D printer shown in panel (**a**) is used to make 3D printed gelatin membranes with gradient configurations suitable for regeneration of heterogenous interface tissues and heterogeneous membrane-shape organs (**b–f**). Panels (**c**) and (**d**) present the first and last layers of a membrane, and panel (**d**) displays a modified configuration with smaller pore size as a result of altering the strand-printing angels. Panel (**f**) presents an SEM image of the membrane. Reproduced from reference [19] with permission

Authors offered that the side with small pore size (first layer) was designed to grow keratinocytes for the formation of epithelial layer. And the side with large pore size (last layer) was designed to accommodate fibroblasts, which are the cells required to make the connective tissue of a full-thickness oral mucosa. Note that having a zero pore size on one side was not acceptable in this membrane, as in most scenarios for growing biological tissues, since the construct for growing a tissue is needed to be permeable for enabling nutrient diffusion and waste removal during the cell growth [19].

7.3   Osteomucosal Tissue Regeneration

Regeneration of bone and oral mucosa has been widely investigated independently. The oral osteomucosal tissue is an interfacial tissue formed by close juxtaposition between the oral mucosa and the underlying bone. Regeneration of oral osteomucosal tissue is helpful for successful intraoral and extraoral grafting, as well as alveolar reconstruction. Osteomucosal constructs could also be used as an alternative model to animal tissues for in vitro drug screening tests. Besides, they could be implemented as study models to investigate the interactions of biomaterials and oral tissues [20, 21].

In a study by Bae et al. [22] osteomucosal constructs were fabricated to be used for regeneration of oral hard and soft tissues. For this purpose, oral fibroblasts and keratinocytes were cultured on an acellular collagen matrix obtained from calvariae of neonatal rats. Histological studies revealed the formation of the osteomucosal constructs composed of a differentiated epithelium and the underlying bone. In addition, differentiated osteocytes and newly formed periosteum-like tissue were observed in the constructs. The structure and architecture of the bioengineered constructs were similar to that of the rat palates with connective and epithelial tissues formed directly above the bone. The osteomucosal tissue constructs are applicable as in vitro models for studying the association of oral soft and hard tissues.

Almela et al. [20] used a biomimetic approach to develop an osteomucosal tissue model by combining the oral mucosa and bone-like structures. They seeded rat osteosarcoma cells on a composite HA/TCP scaffold and cultured the constructs in bioreactor for 3 months. To fabricate the bioengineered oral mucosa, OKF6/TERET-2 oral keratinocytes were seeded on collagen gel-embedded fibroblasts. A biocompatible fibrin-based adhesive was used to combine the engineered bone and the oral mucosa model. Results showed that cellular viability was maintained during the experiments. Besides, histological examinations revealed that the cells were densely proliferated in scaffold pores. In addition, parakeratinized epithelium and connective tissue layer with regularly distributed fibroblasts and widely scattered osteosarcoma cells were observed within the bone scaffold (Fig. 7.2). This study

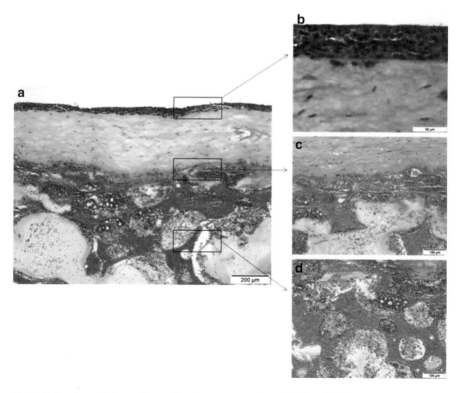

Fig. 7.2 Sections of the engineered oral osteomucosal model (**a**); epithelial (**b**); connective (**c**); and hard tissue layers (**d**). Reproduced from reference [20] which is distributed under the terms of the Creative Commons Attribution 4.0 International License (http://creativecommons.org/licenses/by/4.0/)

showed that fabrication of oral osteomucosal tissue models with anatomical structure and organization is feasible.

7.3.1 A 3D Printed Osteomucosal Model and Mechanism of Adhesion between Hard and Soft Tissues Sections

We have recently proposed a design to develop an integrated 3D human osteomucosal model using 3D printing technique. The project was meant to address the challenging problem of creating multi-tissue constructs composed of two widely dissimilar tissues. The osteomucosal complex is used as an example of constructs composed of different tissue types.

The design took advantage of the capability of 3D printing in making gradient scaffolds. As shown in Fig. 7.3, a gradient 3D printed design is considered for developing the osteomucosal scaffolds. This design allows the allocation of specific cells into the desired areas.

Different compositions of hard materials such as polycaprolactone (PCL) and tricalcium phosphate (TCP) can be used for the hard part. Fibroblast-incorporated collagen can also be employed for developing the soft connective tissue compartment. Then, osteoblasts encapsulated in collagen can be embedded into the hard scaffold, following by incubating the designed bone/connective tissue construct in a bioreactor. To make the epithelial layer, oral keratinocytes can be seeded on top of the construct, and should be cultured at an air–liquid interface to produce a full-thickness oral mucosa on top of the bone section.

Obtaining strong interfacial adhesion has been a challenge with multi-tissue scaffold designs. A number of adhesives can be employed to combine reconstructed soft and hard tissues, which can be classified into two main categories: synthetic/semi-synthetic and biologically derived adhesives [6, 23, 24]. An optimal adhesive should not only be biocompatible but also should be able to hold the two surfaces together with sufficient mechanical strength. In addition, they must be functional in in vitro and in vivo environments. Fibrin glues and collagen adhesives are two main examples of biologically derived adhesives, which are effective in some applications [25]. However, since they originated from human or animal tissues, they are rather expensive and have limited availability [26, 27]. Synthetic and semi-synthetic adhesives such as gelatin–resorcinol–formaldehyde and cyanoacrylate have also been employed over the past years. These adhesives have some complications including limited biodegradability, low biocompatibility, and low adherence to wet surfaces [28, 29].

Using physical adhesion methods that combine either heat or solvent treatment with press-fitting restricts the ability for cell growth in the interfacial region [2]. Although the use of biocompatible natural adhesives, such as fibrin glue, maintains scaffold properties to some extent at the interface to ultimately allow for cellular integration between layers, it hampers direct interfacial integration to achieve a desired interfacial region.

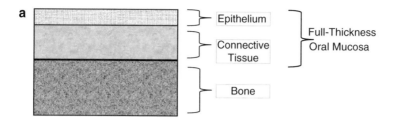

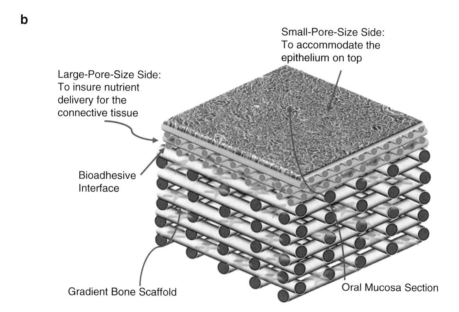

Fig. 7.3 (**a**) The structure of the full-thickness oral mucosa and the underlying bone. (**b**) The scaffold of the proposed 3D printed osteomucosal model

The approach for the adhesion of soft and hard sections suggested for the design of Fig. 7.3 is a biocompatible chemical procedure based on Michael reaction [30, 31]. More specifically, as shown in Fig. 7.4, the soft section can be treated by tyrosinase and the hard section can be treated by 1, 6-hexanediamine. Amino groups are covalently introduced onto the surface of the polymer containing the hard construct (e.g., PCL) by the reaction between 1, 6-hexanediamine and the ester groups of the polymer to obtain the aminolyzed material. One amino group ($-NH_2$) of hexanediamine reacts with the ester group ($-COO-$) of the polymer to form an amide linkage ($-CONH-$), leaving the other amino group available for subsequent reaction. By use of the NH_2 groups as active sites, biocompatible macromolecules, such as collagen, can further be attached to the aminolyzed polymer via a cross-linking agent like tyrosinase [31].

Tyrosinases are binuclear copper-containing enzymes that catalyze the oxidation of phenols in tyrosine residues of hydrogels, such as collagen, into reactive

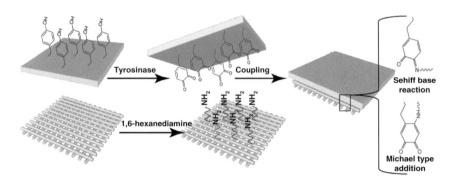

Fig. 7.4 Schematic of the mechanism of adhesion

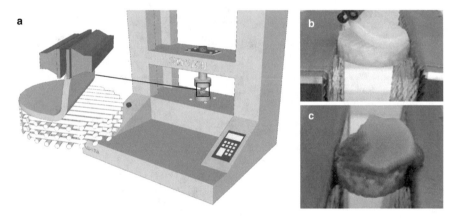

Fig. 7.5 (**a**) Adhesion test setup, (**b**) untreated sample: poor adhesion (adhesion < cohesion), (**c**) treated sample: Strong adhesion (Adhesion > Cohesion) [194]

o-quinones (diphenolase activity). Tyrosinases introduce 0.5 equivalents of dioxygen into their substrates (e.g., tyrosyl residues in collagen) leading to quinones, while 0.5 equivalents of dioxygen are reduced to water. Quinones can take part in Michael additions with amino or hydroxyl groups to form the cross-links. In other words, the interfacial integration between soft and hard sections can be induced through tyrosinase-triggered formation of covalent bonds between o-quinone in the soft section and available amine on the hard surface.

The adhesion test setup is shown in Fig. 7.5a. Alteration from the developed adhesion protocol described above can result in the poor attachment of soft/hard tissues. For example, as shown in Fig. 7.5b and c, while without the incorporation of collagen-containing tyrosinase, we did not have good adhesion (Fig. 7.5b). Such incorporation resulted in strong adhesion of the soft and hard constructs (Fig. 7.5c). More specifically, the cell-seeded soft hydrogel (e.g., collagen) part of the construct is strongly adhered to the hard ceramic/polymer scaffolds. As understood from Fig. 7.5c, the adhesion strength is much greater than the cohesive strength. This means that force applied from the top to detach the soft tissue from the hard construct can rupture the soft part, and actually split the molecules within the soft tissue (cohesive strength) without being able to separate the soft tissue from the hard one.

This method of adhesion, after proper optimizations, can have three main advantages: (1) It is an in situ biocompatible chemical process and does not involve a third material, such as glue, that may affect the tissue growth; (2) The adhesion strength can be high and not diminish over time; (3) The adhesion formulation can lead to enhanced cell growth inside the scaffolds.

In this design, the 3D printing technique provides an in situ encapsulation of the fibroblasts into the matrix with predesigned microstructure, which cannot be achieved with other fabrication methods. Difficulties in migration and penetration of fibroblasts into oral mucosa scaffolds are a serious issue that often can cause the failure of oral mucosa tissue-engineered constructs [32, 33]. Thus, 3D printing offers a relevant and effective solution for this problem by the cell-laden bioprinting technique.

7.4   Osteochondral Tissue Regeneration

Osteochondral defects refer to the cartilage damage accompanying the injury of the adjacent subchondral bone. Osteochondral interface defects in oral and maxillofacial surgery (e.g., mandibular condyle), similar to osteochondral defects in other parts of our body, are known to be highly challenging defects to treat.

Even though significant progress has been made to handle such defects, none of the developed therapeutic methods have yet been proven to ensure long-lasting regeneration. Some treatment methods are also limited due to the lack of osteochondral graft supplies. Therefore, researchers are looking for new approaches that provide long-lasting regeneration to avoid joint replacement and subsequent postoperative complications. A tissue-engineering method using different kinds of scaffolds is one of those developing approaches.

Efforts in the generation of osteochondral tissue engineering scaffold have been made using a wide variety of strategies. One of the most promising is the fabrication of heterogeneous scaffolds, obtained by the combination of distinct but integrated layers corresponding to the cartilage and bone. Such design is based on the recognition of the different requirements to regenerate the cartilage and bone within an osteochondral defect, and at the same time, prevent the risk of delamination of different components if these are adjacent but physically separated.

Among different scaffold requirements to regenerate the cartilage and bone portions of an osteochondral defect in the corresponding regions, pore size, and surface characteristics are highly effective. It has been frequently reported that each cell type requires a certain pore size range in order to obtain the optimum regeneration results [34–36]. Scaffolds with the optimum pore size for each cell type provide an appropriate substrate for cell growth, since they establish a balance between the advantages and disadvantages of varying the scaffold's pore size. For instance, large pore size or porosity of the scaffold may allow for effective nutrient supply, gas diffusion, and metabolic waste removal but lead to low cell attachment and intracellular signaling, while small pore size or porosity can produce an opposite cellular response [35–37].

Thus, employing the 3D printing method can be highly useful, as it can make distinct but integrated layers corresponding to cartilage and bone regions to provide appropriate environment for regeneration of each segment. Other methods were also used in making such scaffolds.

To be more specific, there are three general types of biomaterial-based scaffolds that have been used for regeneration of osteochondral defects [38]:

1. Monophasic scaffolds, in which one material with homogenous porosity is used.
2. Biphasic scaffolds, in which either two different materials or one material consisting of two parts with different porosities are used.
3. Triphasic or multiphasic scaffolds, are scaffolds with three or more materials, or one material consisting of three or more parts with different porosities are employed [38].

As described earlier, biphasic and multiphasic osteochondral scaffolds are more effective for cartilage regeneration and can be made with both natural and synthetic materials [38–42].

The materials selected for different phases of the osteochondral constructs depend on the fabrication method of the construct [43]. The other factor in material selection is the approach for incorporating the cells into the scaffolds [43]. For example, if the intended approach is encapsulation of cells within the scaffolds during the fabrication, the material selection is different than a scenario in which the cells are seeded into the scaffolds after fabrication [43].

One of the approaches in making such bi/multiphasic osteochondral scaffolds is fabricating separate bone and cartilage parts, then connecting them together by different methods, such as press fitting, suturing, or using a glue [45]. For example, in scaffolds made of two phases of fibrin/PCL or PCL/PCL-TCP (Fig. 7.6), fibrin glue

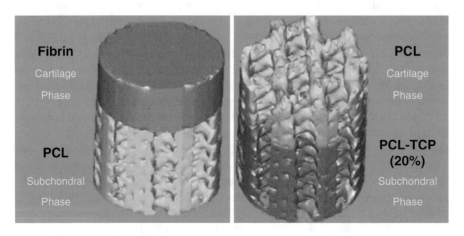

Fig. 7.6 Biphasic osteochondral scaffolds made of fibrin/PCL (left) and PCL/PCL-TCP (right). Reproduced from reference [44] with permission

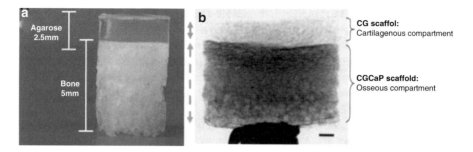

Fig. 7.7 (**a**) A biphasic osteochondral scaffold made of agarose gel and decellularized bone for the cartilage and bone parts, respectively. Reproduced from reference [46] with permission. (**b**) A multiphasic osteochondral scaffold made of collagen type I/glycosaminoglycan/calcium phosphate (CGCaP) for the bone part and collagen type II/glycosaminoglycan (CG) for the cartilage part (Micro-CT image). The scale bar is 1 mm. Reproduced from reference [49] with permission

was used to connect the phases after seeding them separately in chondrogenic (for cartilage regeneration) and osteogenic (for developing bone) media [44, 46]. There are reports indicating that such methods can result in poor connection between cartilage and bone parts and ultimately lead to dissociation [47].

In another biphasic scaffold, agarose gel and decellularized bone were used to make the cartilage and bone parts, respectively (Fig. 7.7a) [46]. Agarose provided the good mechanical property for immature chondrocytes [46, 48], while the decellularized bone was helpful in terms of making osteoinductive construct with mechanical properties and biochemical composition similar to bone [46]. Five hundred micrometers of agarose layer penetrated into the bone portion and solidified at room temperature to form an interface between the two cartilage and bone segments. However, the ultimate interfacial tissue after incubation in an osteochondral bioreactor was distinctive from the interfaces of a native tissue [46].

Attempting to make multiphasic osteochondral scaffolds, Harley et al. used sequences of collagen-type I/glycosaminoglycan/calcium phosphate (CGCaP) in scaffolds fabricated by employing the freeze-drying technique [50]. In another study by the same research group, they made a specific multiphasic scaffold using a mineralized CGCaP and an unmineralized collagen type II/glycosaminoglycan (CG) suspensions for osteochondral regeneration (Fig. 7.7b) [49]. In this scaffold, the interface of the two parts was formed by interdiffusion among the suspensions of each layer before freeze-drying [49]. The authors did not investigate the cell proliferation in the fabricated scaffolds.

The method of interfusion to form an interface between phases of scaffolds was also used by Wang et al., where they employed articular cartilage ECM (ACECM) for cartilage regeneration and HA for bone regeneration in a biphasic osteochondral scaffold [51]. It was evident that the cartilage portion could accommodate the rabbit chondrocytes well, yet not many cells were observed at the interfacial region. The fact that no chondrocytes migrated into the bony portion demonstrated the barrier characteristics of the interface [51].

A grouping of ß-TCP blocks and scaffold-free sheet of mesenchymal stem cell (MSCs) was used by Miyagi et al. for osteochondral applications [52]. In a similar strategy, centrifuged chondrocyte cell sheets were employed by Niyama et al. [53]. However, it was found that using such cell-sheet method has some technical difficulties and limitations, which are mostly related to the use of one kind of cell culture medium in a cell-sheet construct to stimulate the differentiation of both osteoblasts and chondrocytes [52, 53].

Tuan et al. fabricated a biphasic scaffold in which the PCL and PCL/TCP were used for the cartilage and bone parts, respectively [54]. To make this scaffold, a combination of FDM 3D printing method and electrospinning was used. The investigators have implanted this scaffold in pigs and achieved successful cartilage regeneration. However, it was found that not only the scaffold design affected the results but also the implantation site had considerable influence (e.g., medial condyle against patellar groove) [54].

A multiphasic scaffold was produced by Jeon et al., in which one part was composed of 2% alginate and the other part was a biphasic scaffold made of PCL [55]. A combination of FDM 3D printing method and electrospinning was used to fabricate this construct [55]. The two phases were press-fitted to allow the alginate to infiltrate into the PCL part. The scaffold was implanted in rats, and the histological analysis showed separation of some alginate from PCL, indicating the failure of the interface [55].

Another biphasic scaffold was developed using freeze-drying technique, in which the cartilage part was made of chitosan/collagen, while the bone part was composed of ß-TCP [56]. A glue composed of cross-linked 1.2% (w/v) sodium alginate and calcium chloride formed the connection between the two parts [56].

Among other bi/multiphasic osteochondral scaffolds, we can refer to PCL/alginate construct [57], tri-layered PEG-based scaffold with varying ECM composition [58], and bi-layered chitosan/gelatin construct [59].

While regeneration of osteochondral defects—such as the ones related to femoral condyle, humeral head, talus, capitulum of the elbow, and knee [60–70]—using tissue engineering approaches has been investigated, tissue engineering treatment of the osteochondral defects in the maxillofacial area, specifically mandibular condyle [71], is much less studied.

Destruction in osteochondral interface at the temporomandibular joint (TMJ) can cause painful problems in daily chewing, yawning, and talking. There are a few studies that examine the tissue-engineering approaches in resolving this problem. Hollister et al. employed image-based (i.e., computed tomography (CT) and magnetic resonance imaging (MRI)) technique to make patient-specific scaffolds to be used for mandibular condyle, orbital floor, or general mandibular defect regeneration [72]. The applications of the fabricated scaffold were examined in vivo [72]. The authors of another paper took advantage of the developed image-based technique, as well as the solid free-form (SFF) fabrication method, to make biphasic osteochondral scaffolds with load-bearing property using poly-L-lactic acid/hydroxyapatite composite [73]. It was shown that the biphasic poly-L-lactic acid/hydroxyapatite composite scaffold can make bone and cartilage tissues in distinct

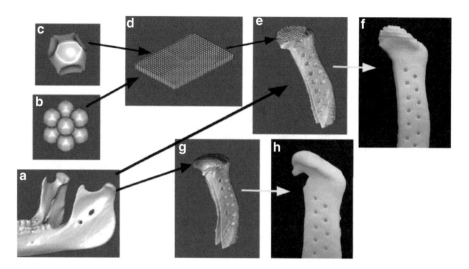

Fig. 7.8 The scaffold preparation for mandibular condyle of Yucatan minipig using the image-based design. Panel (**a**) presents the condyle CT scan and panels (**b**)–(**d**) show the porous construction databases according to the bone and cartilage geometrical equations. The design of the mandibular condyle scaffold with external anatomical shape, surgical fixation, and inner porous construction is shown in panel (**e**). The actual PCL-made scaffold of mandibular condyle is displayed in panel (**f**). The final scheme of mandibular condyle scaffold, representing anatomical shape, surgical fixation, and shell condylar area, is presented in panel (**g**). The actual PCL-made mandibular condyle scaffold shell design is shown in panel (**h**). Reproduced from reference [77] with permission

areas of the scaffolds along with stable interface between them, to be used for TMJ joint repair [74]. Using selective laser-sintering (SLS) technique, a prototype of mandibular pig condyle was developed made of polycaprolactone (PCL) seeded with bone morphogenetic protein-7 (BMP-7) to potentially used in bone/cartilage repair [75]. Manufacturing of such scaffolds with pore sizes of 300–1200 μm were examined using different degradable polymers, calcium phosphate ceramics, and titanium. Depending on the materials and the pore size, the compressive modulus values of the scaffolds were in the range of 50–2900 MPa, while compressive strength values were in the range of 2–56 MPa. The in vivo examinations of such scaffolds in a mouse model indicated the appropriate bone ingrowth in such 3D printed scaffolds [76].

In another study, using the developed images-based design and SLS method, the reconstruction of TMJ was investigated with more detail [77]. Figure 7.8 shows the details of design and fabrication steps. The condylar ramus unit (CRU) scaffold was made of PCL in solid cylindrical shape [77]. The compressive modulus and compressive strength values were 122 MPa and 11.7 MPa, respectively. Comparing these values with human mandibular trabecular bone, they are at the lower end of the mechanical properties for mandibular trabecular bone in the condyle area. The values of compressive modulus and a compressive strength of similar SLS scaffolds with 50% porosity were reported to be 55 MPa and 2.3 MPa, respectively [76, 78].

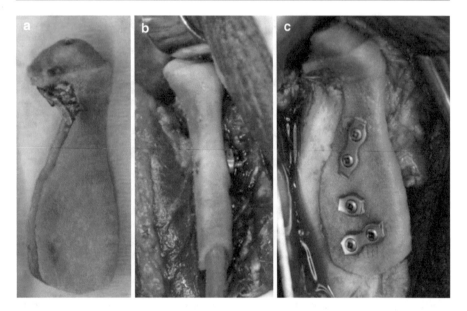

Fig. 7.9 As shown in panel A, the head of the mandibular condyle scaffold was packed with Iliac crest bone marrow. Panel B shows the good adaptation of the scaffold into the native mandible and panel C shows how the scaffold is secured to ramus. Reproduced from reference [77] with permission

The fabricated scaffolds were applied for reconstruction of TMJ in a Yucatan mini-pig animal model, while autologous iliac crest bone marrow was packed in the condylar heads of the scaffolds, as shown in Fig. 7.9. To investigate the efficacy of the scaffold, micro-CT and histology analyses have been performed; it was shown that bone volumes and tissue mineral density were significantly increased after 1 and 3 months, and new bone was grown at the interior and exterior of the scaffold (Fig. 7.10). Growth of cartilage-like tissue was also observed after 3 months [77].

7.5   Periodontal Tissue Regeneration

Periodontal tissue, also called periodontium, is a complex organ composed of root cementum, periodontal ligament (PDL), alveolar bone, and gingiva. Each of the periodontium constituents has distinctive biochemical, cellular, and architectural properties. Basically, a periodontal defect is known as an example of complex multi-tissue defect of combined soft (gingival and PDL) and hard (bone and cementum) tissues that can be caused either by disease (such as periodontitis) or trauma. Such periodontal defects may eventually cause tooth loss if left untreated.

The periodontium acts as a structural support for the teeth and adsorbs the mechanical shocks that occur during mastication by providing a resilient attachment for the teeth within the alveolar sockets. Another critical role of the periodontium is that it provides a cellular reservoir for dynamic tissue remodeling and repair [79].

Fig. 7.10 The micro-CT images indicate the formation of the new bone above the condylectomy cut at the interior (orange) and exterior (yellow) of the scaffold. Reproduced from reference [77] with permission

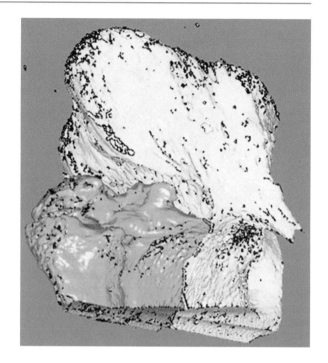

Periodontal problems are highly widespread and about 46% of the population are influenced by it [80]. However, unfortunately, the treatment options are not robust, and the outcomes are not predictable, especially in case of complete periodontal regeneration. The origin of the challenge can be due to: (1) the complexity in the architecture of multi-tissue periodontium, (2) shortage of soft tissue in some defects, (3) the fact that different tissues must be healed in a spatiotemporally coordinated manner, (4) the avascular nature of tooth surface, and (5) the susceptibility of the wound to infection due to the presence of intraoral bacteria [81].

Periodontal defects, in the majority of the cases, are the result of the inflammatory response to a bacterial infection. The accumulation of a bacterial biofilm at the dental surfaces triggers inflammatory responses that result in the progressive loss of periodontal tissues. Uncontrolled loss of periodontal tissues can contribute to tooth loss. Furthermore, it has been demonstrated that periodontal diseases can cause serious risks for general health [82, 83]. When the inflammation is limited to the gingival tissue only, the damage is reversible. When all periodontal tissues, including bone, cement, periodontal ligament, and gingiva, are involved in the inflammation, the damage is currently considered irreversible. The kind of bone defect determines the possible treatment options and their outcome. Bone loss can be characterized as either horizontal (flat) or vertical (crater-like) bone loss. Bone loss also results in the recession of the covering soft tissue (gingival recessions). Vertical bone loss, or intrabony defects, can have different architectural characteristics. The bone defect at a tooth can be well-defined with three surrounding bone walls or less defined with only one or two surrounding walls present. Figure 7.11 shows a regular case of a periodontal defect that might be caused by bacterial biofilm.

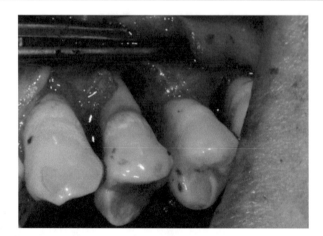

Fig. 7.11 A gingival flap exposes the intrabony defects distal of tooth #12 and mesial of #14. The vertical bone defect distal of tooth #12 (first premolar) can be characterized as a two-wall defect. The defect is open to the buccal site but surrounded by bone walls toward the adjacent tooth and the palate. The vertical defect at tooth #14 (first molar) is well surrounded by bone and can be characterized as a three-wall defect. Be aware that verticals defects often also have a horizontal component. That can be seen in comparing the bone heights of the first and second premolars. The mesial bone height at the first premolar is higher than at the second premolar

The key characteristics of an advanced periodontal defect are illustrated in Fig. 7.12. The periodontal lesion is characterized by the presence of a gingival recession (loss of gum tissue), the development of a periodontal pocket, the breakdown of connective tissue and the periodontal ligament, and the resorption of bone, with a loss of bone height.

Please note that gingival recessions can also have non-inflammatory causes. The gingival margin is apical of the Cementum-Enamel-Junction (CEJ) and leaves the root surfaces that are covered with cementum exposed. Figure 7.12 shows that in healthy periodontium, the bone level is approximately 2 mm apical of the CEJ. When the gingival margin is apical of the CEJ, the bone level is also reduced.

The majority of treatment goals in periodontal therapy are the resolution of the inflammation, which is caused by bacterial infection, and the regeneration of periodontal tissue that was lost to the inflammatory destructive processes. Since complete periodontal regeneration is currently not possible, resective treatments that aim at correcting of periodontal defects and anatomical problems are conventionally used to treat patients with horizontal bone loss [84].

The comprehensive treatment of a patient with periodontitis is complex but can be categorized into three main steps. (1) The non-surgical treatment that aims to stop the periodontal infection and the responding inflammation. Clinical outcomes are a reduction of pocket depths and clinical signs of inflammation, like bleeding on probing. (2) Residual pocket depths and intrabony defects are treated in the surgical-corrective phase with regenerative or resective techniques. (3) The treatment outcome is maintained by regular and frequent supportive-periodontal treatments (maintenance phase) [85].

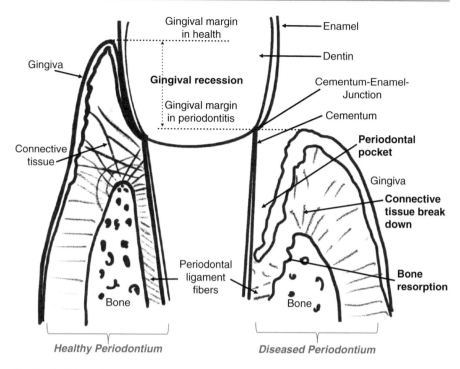

Fig. 7.12 Schematic comparison of a healthy periodontium (left hand) and a periodontal defect (right hand) as a result of periodontitis. The gingival tissue attaches to the Cementum-Enamel-Junction at the tooth. The tooth is connected to the bone via periodontal ligament that connects cementum with bone. In a periodontal defect, the gingival margin follows the bone level. The inflammatory processes lead to the destruction of all compartments of the periodontium

Clinicians can choose between different treatment options for the regeneration of periodontal tissue. The use of biologicals like enamel matrix derivative (EMD) and recombinant human platelet-derived growth factor-BB (rhPDGF-BB) shows improvements in clinical parameters that are comparable with that of bone replacement grafts and GTR [86]. The clinical outcomes of these treatments are a gain in clinical attachment, decreased pocket probing depth, gain in radiographic bone height, and overall improvement in periodontal health. The regeneration is most effective when applied in early lesions without or with a limited presence of gingival recessions [87].

As it will be described in detail during the following sections, the principle of GTR is based on the hypothesis that epithelial cells and fibroblasts can be prevented from migrating into a wound by means of a membrane barrier, while simultaneously providing the space for those particular cells to repopulate the wound, which has the capacity to regenerate the desired type of tissue [86].

In such cases, a periodontist will open a flap and remove the destructed tissue, which can be part or all of the four tissue types of periodontium (gingiva, bone, PDL, and cementum), followed by filling the defect with a bone graft and covering it with a GTR membrane.

Although the GTR surgical technique is now very common in periodontal regeneration, it is reported to be highly unpredictable and dependable on the dentists' skill [88–90]. The clinical outcome depends on multiple factors. Deeper and narrow defects have a more favorable regeneration outcome than wide and flat defects [89]. The more bone walls that surround the defect has the better the response to regenerative treatment [90]. However, the regeneration of intrabony defects is limited by the horizontal level of bone and will not exceed this level. Moreover, there are many statements about variability in the clinical efficacy of the technique [91]. It was concluded by the investigators that the lack of complete compartmentalization between the different parts of the multi-tissue defects during the treatment can be responsible for the chaos in the clinical outcomes, and GTR may not completely provide the compartmentalization that is required for the regeneration of such tissues. More specifically, each of the four tissues requires their own cell accommodation and specific environment to be grown, which cannot be easily provided in current treatment method.

Here, the interfacial and multi-tissue engineering techniques highlight their position.

7.5.1   Periodontal Regeneration Via Tissue Engineering and 3D Printing Approaches

Conventional treatment of periodontal diseases involves stopping the progression of the disease by eliminating the biofilm through systematic or local administration of antibiotics. However, antibiotic therapy does not lead to the regeneration of the lost periodontal tissues. A successful therapeutic solution should be able to regenerate cementum, periodontal ligaments, and alveolar bone. It should also result in the re-attachment of the epithelial tissues and the re-establishment of the normal periodontal structures [92].

It is hopeful that using regenerative treatments in adjunct to conventional therapies is useful to restore both the structure and function of the destructed periodontium. Tissue engineering is considered a promising candidate to regenerate periodontal structures. The possibility of regenerating periodontium through tissue engineering was suggested for the first time in 1993 [93]. Typically, the following tissues are involved in periodontal regeneration: cementum, PDL, alveolar bone, gingival epithelium, gingival connective tissue, and the periodontal vascular network [94]. To resemble the multi-tissue organization of the periodontium, a complex combination of biomaterials and cells is required in one construct.

More specifically, we need a model that can regain the hierarchically organized network comprised of four tissues of 1-gingiva, 2-alveolar bone, 3-PDL, and 4-cementum. Such model has not been yet developed; however, there are good progress in the field of multi-tissue regeneration that can facilitate the assembly of such model.

Various studies have been performed on development of multi-layered materials for the reconstruction of bone, cementum, and PDL [95, 96], but not gingiva as part of periodontium. More specifically, so far composite one-layer [97], bi-layer [98], and tri-layer [99, 100] have been fabricated. The design and development of a quad-layer construct still lack in this field. We suggest use of the model illustrated in Fig. 7.13, which is basically built upon the described osteomucosal model mentioned earlier in this chapter. The compartmentalized structure of this model follows the complex architecture of a multi-tissue periodontium by considering all four periodontium tissues and may address the challenging and unresolved problem in defects that suffer from shortage of soft tissue as well.

As described earlier, periodontal defects can have different architectures. Due to variations of defect kinds and shapes in different clinical cases, we suggest this model be built in two different assemblies, sided configuration (Fig. 7.13a) and layered configuration (Fig. 7.13b), to provide the opportunity for surgeons to use the construct based on the characteristics of clinical cases. For example, the sided configuration can be useful for the clinical cases of an intrabony defect, and the layered configuration is more suitable for the rescission cases, in which this model is aimed to regain the bone height, regenerate the periodontal ligament and cementum, as well as regenerate the lost gingival tissue to establish a gingival margin to a position that is similar to that of healthy periodontal tissue.

Of course, the capability of 3D printing in making constructs for the regeneration of heterogenous tissues is useful in the realization of such models. For example, the bone/PDL section of this model has been prepared by our group as described below.

Figure 7.14 shows the design of the multilayered membrane for regeneration of the Bone/PDL interface. It is composed of a supportive layer, which is a 3D printed mesh for the bone regeneration on one side. On top of this layer, a freeze-dried porous hydrogel is designed to facilitate the PDL regeneration. The 3D printed

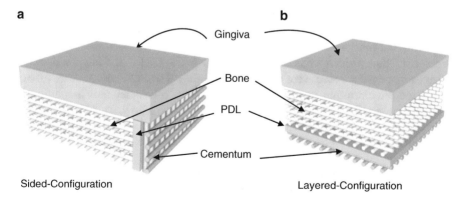

Fig. 7.13. (**a**, **b**) A quad-layer model for the reconstruction of Gingiva/Bone/PDL/Cementum tissues in two different configurations, the sided-configuration and the layered-configuration, to cover the various architectures of the periodontal defects in different clinical cases. For example, the sided configuration can be used for the case of intrabony defects, and the layered configuration is suitable for the recession cases. 3D printing can provide good opportunities for realizations of such models

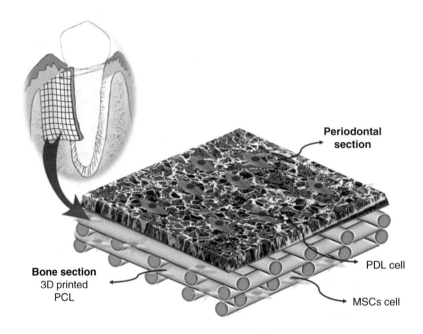

Fig. 7.14 Design of a 3D printed multilayered membrane for regeneration of PDL-bone interface [31]

mesh can also act as a supportive layer to improve the mechanical property of the entire construct [31].

Figure 7.15 presents the actual image of the fabricated multilayered membrane in which the 3D printed mesh was made of PCL using hot-melt extrusion by using a commercial bioprinting instrument (3D-Bioplotter® Manufacturer Series, Envisiontec, Germany). Using a nozzle with 0.4 mm inner diameter, a $15 \times 15 \times 0.2$ mm^3 mesh was printed, while the distance between strands was considered 0.5 mm with an angle of 90° shift (Fig. 7.15a) [31]. A 2.5 bar pressure and 4 mm/s dispensing speed were set for this printing. The PCL construct was functionalized after printing by immersing in 5 wt% 1, 6-hexanediamine solution in isopropanol for 1 h at 37 °C to graft 1, 6-hexanediamine onto them. Following rinsing and immersing in deionized water, hydrocaffeic acid, 1-ethyl-3-(3-dimethyl-aminopropyl) carbodiimide (EDC), and ethanol were included in the water [31]. After necessary incubation and adjustment of pH, the resulting treated PCL was called modified PCL (MPCL) (Fig. 7.15b) [31]. It was claimed that the resulting functionalities on the PCL membrane make connection with nucleophiles, such as amine groups, which suggests improved protein condensation and cellular attachment.

A collagen solution was freeze-dried on top of the MPCL 3D printed membrane [31]. Following the cross-linking of the collagen porous matrix with EDC, bone morphogenic protein (BMP-2) was immobilized on it [31].

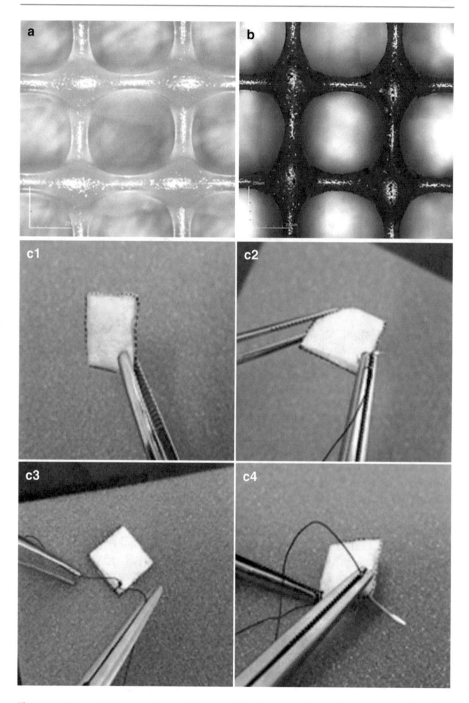

Fig. 7.15 The actual image of the multilayered 3D printed membrane for the regeneration of the PDL/Bone interface: (**a**) The 3-printed PCL membrane, (**b**) the modified PCL (MPCL) mesh, (**c1**) The prepared construct where the collagen porous matrix is fabricated on top of the MPCL membrane. (**c2–c4**) The suturability of the multilayered membrane [31]

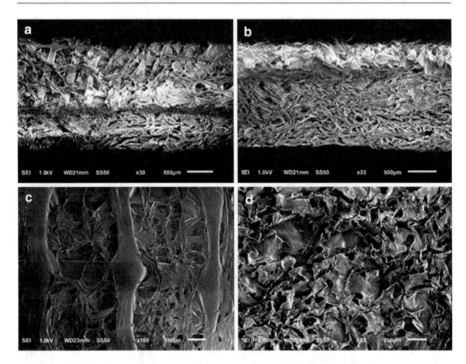

Fig. 7.16 SEM images of the 3D printed multilayered membrane: Panel (**a**) presents a tilted view and panel (**b**) shows the cross-sectional view. Panel (**c**) is the image of the membrane from the side with MPCL mesh for bone regeneration and panel (**d**) displays the side of the membrane covered with collagen porous matrix for PDL regeneration [31]

Figure 7.15c1–c4 presents the actual image of the membrane after preparation, in which its capability to hold sutures display its high-suture retention strength, suggesting easy surgical handling of the fabricated scaffold. Scanning electron microscopy (SEM) images of the construct are shown in Fig. 7.16 that illustrates the secure adhesion between the bone and PDL sections as a result of functionalizing the PCL membrane [31].

The observation of PDL cells which were well spread onto the membrane and did not infiltrate into the bone section of the membrane (Fig. 7.17), suggests the potential applications of the developed construct for GBR [31], as it will be described in the following section.

Combing this 3D printed Bone/PDL construct and the 3D printed osteomucosal model described earlier in the chapter can be a good pathway for developing the model shown in Fig. 7.13 for complete multi-tissue periodontal regeneration.

PDL fibroblasts and MSCs were cocultured onto the fabricated membrane presented in Fig. 7.15. Figures 7.17 and 7.18 show the images and in vitro analyses of the cells on each side of the membrane, suggesting that MPCL and collagen can make appropriate environments for the growth of bone and PDL, respectively.

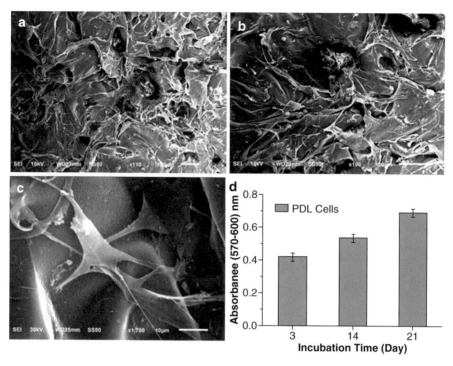

Fig. 7.17 The collagen side of the membrane with attached PDL fibroblasts (SEM images) is shown in panels (**a–c**). The in vitro analysis presented in panel (**d**) indicated the good proliferation rate of cells and its appropriateness for PDL regeneration [31]

Moreover, the in vitro analyses presented in Fig. 7.18 indicated that the modification and functionalization of PCL facilitate the cell attachment, proliferation, and general accommodation of MSCs.

7.5.2   Periodontal Tissue Regeneration Via Cell Sheet Engineering

In conventional tissue engineering approaches, cells are released by performing ECM fragmentation. The proteolytic enzymes used in this process could hydrolyze the membrane proteins and damage the cells. A solution for this problem is cell sheet engineering. This technique eliminates the need for proteolytic enzymes and provides an efficient way to harvest intact sheets of cells. Since the cell–cell and cell–ECM connections are kept intact, the cells' survival rate is not altered after transfer. The harvested sheets can be transplanted directly to the implantation site. It is also possible to overlap the sheets and create three-dimensional structures [101, 102].

In cell sheet technique, adhesion and detachment of the cells on the culture dish are controlled by temperature changes. Cells are cultured on the surface of a

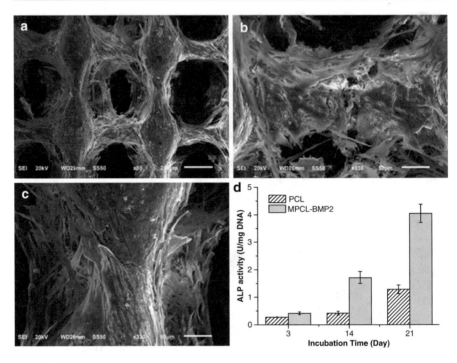

Fig. 7.18 SEM images from the differentiated MSCs after 21 days culturing on the bone side of the membrane are presented in panels (**a**)–(**c**). Panel (**d**) compares the ALP activity of MSCs (during the 21 days) on PCL and MPCL-BMP2 using ALP and DNA Pico green assays [31]

thermo-responsive polymer, such as poly(N-isopropylacrylamide) (PIPAAm). At temperatures lower than 32 °C, PIPAAm is hydrophobic and allows the cells to adhere to its surface. At temperatures higher than 32 °C, PIPAAm turns into a hydrophilic polymer and the cells detach from its surface [103].

Cell sheet technique has been increasingly used for regenerative purposes, including periodontal regeneration, myocardial regeneration, and cornea transplantation [104–106]. Due to the complex and multi-tissue structure of the periodontium, the regeneration strategies based on single MSC types may not accomplish the requirements of periodontal regeneration.

In 2016, Zhang et al. constructed a cell sheet composed of PDLSCs and BMMSCs derived from the jaw bone (JBMMSCs) [107]. They cocultured PDLSCs and JBMMSCs and found crosstalk between these MSCs. As a result, the expression of bone and ECM-related genes improved significantly in both MSCs, with the resultant composite structure being similar to the natural periodontal tissue. Transplantation of the cell sheets in 6-week-old immunodeficient mice led to the formation of PDL/bone-like and PDL/cementum-like structures after 6 weeks (Fig. 7.19). The results showed that the interaction of the cocultured MSCs creates a suitable microenvironment that allows for the construction of tissue structures resembling the native periodontium. The suggested method provides a promising

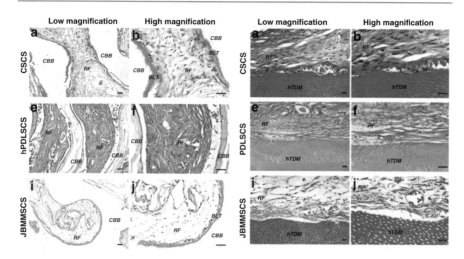

Fig. 7.19 Formation of PDL/bone-like (left) and PDL/cementum-like tissues (right). More periodontal-like tissues are observed in the composite stem cell sheet (CSCS) group. Reproduced from reference [107] which is distributed under the terms of the Creative Commons Attribution 4.0 International License (http://creativecommons.org/licenses/by/4.0/)

approach for the regeneration of functional tissues applicable in the clinical reconstruction of periodontal defects.

In a clinical study by Iwata et al. in 2018, cell sheet engineering was used to regenerate periodontal defects in ten patients with severe periodontitis [108]. PDLSCs were scraped from the surface of extracted third molar of the patients (Fig. 7.20). Cell sheets prepared from autologous PDLSCs were transplanted on the root surface following flap surgeries. The bony defects were filled with β-tricalcium phosphate granules (Fig. 7.21). After 3 and 6 months, CBCT imaging and clinical evaluations were performed. Follow-up examinations were carried out upto 7 years. In all patients, periodontal probing depth and clinical attachment level were reduced significantly. Additionally, CBCT scans acquired after 6 months revealed a significant increase in bone height. Follow-up studies revealed the maintenance of the therapeutic outcomes, with no complications observed in the patients. This study proved the safety and efficacy of PDL cell sheet engineering as a promising strategy to regenerate severe periodontal defects.

In a recent research by Raju et al., a three-dimensional anatomical bone–ligament structure was created via cell sheet engineering [109]. For this purpose, PDL and osteoblast-like cells were deposited on culture dishes coated by PIPAAm. The complex cell sheet structure anatomically reconstructed the bone–ligament tissue after ectopic and orthotopic transplantation in female mice with severe combined immunodeficiency. Eight weeks after implantation, the bone volume to tissue volume ratio (BV/TV) was significantly higher in complex cell sheet group compared to the control and PDL cell sheet group. The functional connections of PDL-like fibers to alveolar bone and teeth root were also regenerated in the complex sheet group (Fig. 7.22). This study demonstrated successful fabrication of a complex

Fig. 7.20 The schematic representation of the procedure for periodontal regeneration using autologous PDL-derived cell sheets and β-tricalcium phosphate granules. Reproduced from reference [108] which is distributed under the terms of the Creative Commons Attribution 4.0 International License (http://creativecommons.org/licenses/by/4.0/)

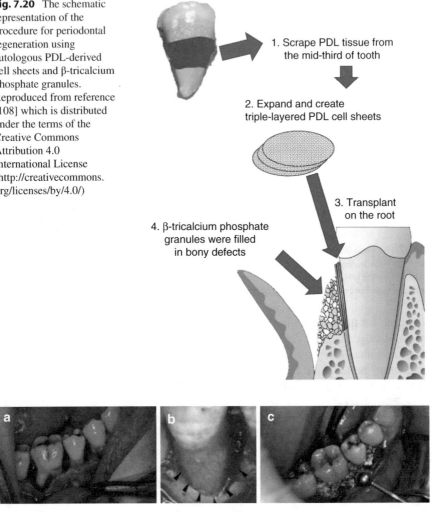

1. Scrape PDL tissue from the mid-third of tooth

2. Expand and create triple-layered PDL cell sheets

3. Transplant on the root

4. β-tricalcium phosphate granules were filled in bony defects

Fig. 7.21 Proposed surgical procedure, including open flap surgery (**a**), transplantation of the PDL-derived cell sheet into the defect site (**b**), and filling of the bony defect with β-tricalcium phosphate granules (**c**). Reproduced from [108] which is distributed under the terms of the Creative Commons Attribution 4.0 International License (http://creativecommons.org/licenses/by/4.0/)

PDL–osteoblast cell sheet that regenerated a large-scale periodontal defect similar to the anatomical structures.

Investigators have also shown that the combination of 3D printing and cell sheet engineering can be an effective approach in periodontal regeneration [2]. More specifically, in a study performed by Vaquette et al. to regenerate the alveolar bone/periodontal ligament complex, a biphasic scaffold was made using the FDM 3D printing method to fabricate a section for bone regeneration, and the electrospinning method was employed to make an electrospun membrane to accommodate the

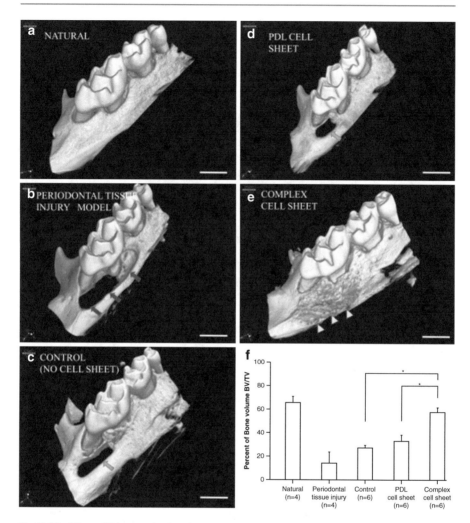

Fig. 7.22 Micro-CT images (**a–e**) and regenerated bone volume in periodontal tissue injury models, 8 weeks after orthotopic transplantation. The scans and data demonstrate the significant difference in the regenerated bone volume of the complex cell sheet group in comparison to the control and the PDL cell sheet group. Reproduced from reference [109] which is distributed under the terms of the Creative Commons Attribution 4.0 International License (http://creativecommons.org/licenses/by/4.0/)

pcriodontal section (Fig. 7.23) [2]. Multiple cell sheets of periodontal ligament (PDL) were placed on the electrospun membrane, while osteoblast cells were cultured on the 3D printed part. The in vitro studies showed that these two parts could make appropriate environments for bone and PDL growth. A successful in vivo study was performed by applying the construct on a dentin block, followed by implanting it in a subcutaneous animal model. Cementum deposition was detected on the surface of the dentin block. In this study, improved attachment to the dentin

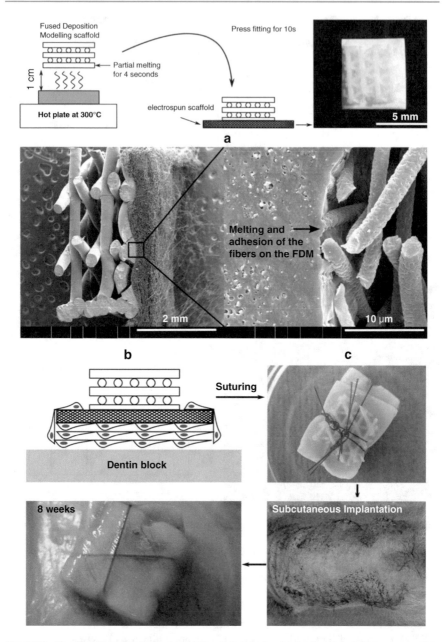

Fig. 7.23 Combination of cell sheet and 3D printing technologies for periodontal regeneration: (**a**) Schematic of the fabrication procedure for making a biphasic scaffold made by 3D printing and electrospinning methods for alveolar bone/periodontal ligament complex regeneration. (**b**, **c**) Cross-sectional view of the scaffold: the 3D printed section and the electrospun section were designed as the bone and periodontal compartments, respectively. (**d**) Multiple cell sheets of PDL are placed on the electrospun periodontal compartments. The resulting hybrid construct was assembled onto a dentin block to be examined in vivo in an athymic rat subcutaneous model. Reproduced from reference [2] with permission

surface was observed in comparison with such attachment using constructs with no cell sheets [2].

7.6   Whole Tooth Regeneration

The whole tooth is considered a multi-tissue organ and making a functional tooth can be a highly complicated process involving several biological factors. We will discuss the whole tooth regeneration in this chapter only because of its multi-tissue nature, although the challenges in its regeneration are more toward the molecular processes, exploring a capable autologous cell source and bioactive materials, rather than issues that can be resolved by 3D printing.

Cooperation of several cells in line with spatiotemporal expression of molecular signals leads to the formation of teeth with different morphologies at their anatomical sites. The proper interaction between these factors is critical to generate functional teeth. Several approaches have been proposed to regenerate whole teeth. The most common approaches include tissue engineering, organ germ method, and cell sheet engineering [110].

7.6.1   Whole Tooth Regeneration Via Tissue Engineering

Tissue engineering employs stem cells, growth factors, and scaffolds to regenerate a whole tooth. Successful regeneration of tooth requires accurate arrangement of odontoblasts, ameloblasts, and cementoblasts. Besides, the selection of proper scaffolds is critical to obtain constructs with desired properties [110].

In tooth engineering, the most commonly used materials for scaffolding are PLGA, PCL, gelatin, chitosan, and collagen. In a study conducted by Chen et al. [111], aligned layers of electrospun PLGA and gelatin were used to regenerate periodontium and dental pulp. Results showed that the composite scaffolds guided cell proliferation and promoted odontogenic differentiation in vitro. The cell-seeded scaffolds were implanted in miniature swine jaws, and 12 weeks after the implantation, tooth-like structures were formed. The authors concluded that the proposed scaffolds exhibit appropriate biocompatibility and physiochemical properties and provide a suitable microenvironment for tooth regeneration.

In another study by Kochler-Bopp et al. [112], PCL scaffolds were used to create bioengineered teeth. PCL was employed to fabricate scaffolds as it holds FDA approval and is capable of mimicking natural ECM. PLGA nanoparticles loaded by immunosuppressive drug Cyclosporine A (CsA) were embedded on the electrospun PCL scaffolds to quicken the innervation of the bioengineered teeth after implantation. Then, the constructs were implanted subcutaneously in adult ICR mice. Histological analyses were performed 2 weeks after implantation. It was demonstrated that the scaffolds did not modify the development of the bioengineered teeth in vivo. Furthermore, innervation of the bioengineered teeth pulp was studied by transmission electron microscopy (TEM) and indirect immunofluorescence. The results revealed that 88.4% of the bioengineered teeth were successfully innervated. This study proved that PCL scaffolds functionalized with CsA-loaded PLGA nanoparticles provide a promising way to develop bioengineered teeth with normal development after implantation.

Another scaffold material for tooth engineering is collagen, which is a natural polymer abundantly found in dental pulp and dentin matrix. Tissue compatibility and antigenic properties make collagen a proper scaffold material. It has been reported that collagen provides a substrate for dentinogenesis by organizing pre-odontoblasts and promoting the adhesion of odontoblasts to the dental pulp. Collagen is an ideal scaffold for pulp tissue engineering due to its differentiative effect on stem/progenitor cells. It has also been demonstrated that collagen hydrogels and sponges promote the differentiation of DPSCs into odontoblasts [113, 114].

Chitosan is known for its extensive use in wound dressings, but it could also be used for pulp capping. In dental tissue engineering, chitosan has been used as a scaffold material due to its inductive effects on attachment, proliferation, and growth of DPSCs [115]. However, it is often necessary to combine chitosan with other biomaterials to improve its bioactivity and mechanical properties. In a study by Chen et al. [116], freeze-dried chitosan/carboxymethyl cellulose (CMC) was used to fabricate scaffolds for pulp tissue engineering. The chitosan/CMC scaffolds exhibited improved internal porosity with reduced pore size. Besides, cellular proliferation and the expression level of osteonectin and dentin sialophosphoprotein were enhanced by the use of composite scaffolds.

In another study [117], dental epithelial cells and DPSCs were seeded on chitosan/collagen scaffolds in order to promote epithelial–mesenchymal interactions. Results showed that the composite scaffold supported the differentiation and migration of dental epithelial cells and DPSCs. The biomimetic structure of the scaffolds also promoted in vivo neovascularization. In addition, calcium ion deposition was observed at the surface of the constructs after 24 days.

7.6.2   Whole Tooth Regeneration Via Organ Germ Method

In the organ germ method, natural organogenesis and organ development are mimicked to induce epithelial–mesenchymal interconnection. In this method, embryonic and non-embryonic epithelial and mesenchymal stem cells derived from tooth germ are used. For whole tooth regeneration using embryonic germ cells, epithelial and mesenchymal stem cells are cocultured in vitro for 5–7 days (Fig. 7.24). This leads to the differentiation of odontoblasts, histogenesis of dental epithelium, and formation of cusp and crown. Then, the construct is implanted subcutaneously or in the sub-renal capsule of the host to form enamel, dentin, and vascular tissues. This approach has been reported to be promising in creating functional bioengineered teeth in the edentulous jaw. However, clinical application of this approach is controversial due to the immunological reactions and ethical issues regarding the use of embryonic cells. These concerns have motivated the search for alternative cell sources [110, 118, 119].

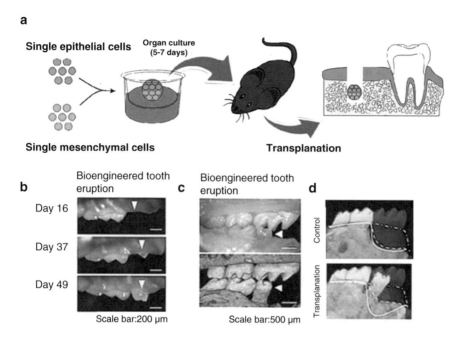

Fig. 7.24 (**a**) Schematic representation of organ germ method to create a bioengineered tooth. (**b**) Transplanted bioengineered tooth prior to (top) and after eruption (middle), and after occlusion (bottom). (**c**) Bioengineered tooth 40 days after transplantation. (**d**) Micro CT images of the control (no transplantation) and transplanted bioengineered tooth after 45 days. Reproduced from [110] with permission

Adult stem cells have been proposed as a potential candidate instead of embryonic stem cells. It has been shown that BMMSCs could be an appropriate alternative for embryonic stem cells applicable in organ germ method. Interconnections between BMMSCs and oral epithelial cells induce the formation of tooth-like structures after being implanted in murine kidney capsules. Another source of adult stem cells for tooth bioengineering is iPSCs. iPSC technologies provide pluripotent stem cells with no ethical or immunological complications. It has been shown that cocultured iPSCs and incisor mesenchymal cells form tooth-like structures [110, 118].

7.6.3 Whole Tooth Regeneration Via Cell Sheet Engineering

During the last two decades, many studies have been focused on whole tooth regeneration. However, despite the great advances in this field, controlling the morphology of the bioengineered teeth remains as an unsolved challenge. Furthermore, the bioengineered teeth are susceptible to eruption after implantation [120].

An alternative approach for tooth regeneration is tooth root engineering. The bioengineered tooth root acts as an anchor for a natural or prosthetic crown. The bioroot complex is constructed by seeding MSCs and MSC sheets on root-like

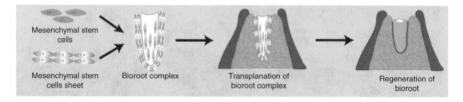

Fig. 7.25 Schematic representation of bioroot regeneration. The bioroot complex composed of root-like scaffold, MSCs, and MSC sheets is implanted into the alveolar bone. Reproduced from reference [120] with permission

scaffolds. The construct is then implanted into the alveolar socket (Fig. 7.25). It has been demonstrated that bioengineered roots exhibit PDL and dentin-like structures, along with providing adequate mechanical support for the prosthetic crown [120].

Cell sheet technology has been increasingly used for the regeneration of dental and oral tissues. It has the potential to establish interfacial epithelial-mesenchymal constructs, which is of great importance for tooth regeneration. It could also be used to fabricate functional bioengineered teeth, should it be implemented in combination with other tissue engineering techniques [110].

It has been reported that DFSCs seeded on dentin matrix scaffolds in combination with autologous fibroblast cell sheets promote the formation of root-like structures [121]. In another study, the surface of DPSCs seeded HA scaffolds was covered by vitamin C-induced PDLSC sheets, and the resulting multilayer construct was transplanted into the alveolar bone socket. After 6 months, the bioengineered root was capable of supporting a prosthetic crown [122].

In a recent study by Meng et al. [123] cell sheet technology was used for tooth root regeneration. For this purpose, a sandwich structure was developed composed of human DPSC sheet, treated dentin matrix (TDM), and Matrigel. After the successful proliferation of DPSCs on TDM and Matrigel, the resulting sandwich structure was transplanted under the skin of nude mice. After 3 months, dentin, pulp, and periodontium-like tissues were regenerated. Furthermore, nerve-like fibers and blood vessel-like structures were formed in the pulp cavity. This study revealed that DPSC sheet along with DPSC/TDM and DPSC/Matrigel could form tissue structures similar to that of tooth root.

7.7   Guided Tissue/Bone Regeneration (GTR/GBR)

The concept of GTR and GBR was initially introduced in 1959 [124] and then recapped more clearly in 1984 [125]. In GTR and GBR techniques, a barrier membrane prevents the migration of cells from the connective tissue or epithelium to the underlying tissues such as bone.

GBR and GTR, as barrier membranes, share similar conceptions and sometimes are used interchangeably, however, GTR typically involves soft tissues and is mostly applies for regeneration of periodontal soft defects [126], while in GBR, the goal is growth of healthy bone in a way that no fibroblast or other interfering cells penetrate into the bony region [127, 128].

The GBR/GTR method is very useful in multi- and interfacial tissue treatment during oral and maxillofacial surgery where the bone (or another tissue) is needed to be separated from a specific fast-growing soft tissue, such as tissues composed of fibroblasts. Without the GBR/GTR barrier membrane, fibroblasts can easily grow and penetrate into a region where another tissue, such as bone, was supposed to grow, and thus make undesired and low-quality target tissues.

A good GTR/GBR membrane must: (1) be biocompatible without prompting inflammatory response, (2) have appropriate configuration and pore size in each side to accommodate relevant and different cell types in different sides of the barrier and to prevent penetration of cells while allowing the nutrient delivery and neovascularization, (3) have proper degradation (if resorbable), (4) have appropriate mechanical properties for secure and easy surgical handling, preventing the membrane collapse, and also maintaining the space (for specific applications) [129, 130].

As we will discuss later, 3D printing provides a very appropriate method for the fabrication of GBR/GTR membranes, due to its nature of layer-by-layer material deposition.

As summarized in Table 7.1, we can categorize the GTR/GBR membranes into two groups of non-resorbable and bioabsorbable membranes. Bioresorbable membranes can be made either by natural or synthetic materials.

7.7.1   Non-resorbable Membrane

The gold standard material for many non-resorbable biomedical membranes is expanded polytetrafluorethylene (ePTFE), which is the porous form of polytetrafluorethylene (PTFE) [131]. PTFE is made of fluorocarbon polymer and is a fully inert and biocompatible material that makes minimal inflammatory reaction in human body [132, 133]. However, since PTFE is nonporous, it prevents tissue ingrowth, thus ePTFE is manufactured with the aim of supplying porosity to the body of PTFE [134, 135]. The structure of the pores in ePTFE has the shape of solid nodes and fibrils, which are formed by expanding the PTFE in a way to create a microporous configuration [134, 136]. The porosity in ePTFE can be variable based on the fabrication condition and parameters. Gore-Tex® is the very first commercialized non-resorbable membrane, which took advantage of this characteristic of ePTFE. In an innovative configuration, Gore-Tex membranes are made with a bilayer configuration, in which each layer has a different porosity (90% and 30%) to deliver a certain function. More specifically, the Gore-Tex membrane is composed of:

1. A membrane with thickness of 1 mm that is 90% porous. It does not allow the migration of apical epithelium cells and facilitates the ingrowth of connective tissue. It is placed coronally and effectively helps the stability of wound [134].
2. A membrane with thickness of 0.15 mm which is 30% porous. This part maintains the required space for regeneration (a space maintainer). The structural

Table 7.1 Summarized information about different types of GTR/GBR membranes

Non-resorbable membranes	
Examples	• GoreTex® made of expanded PTFE (ePTFE) [185]
	• TefGen-FD® made of nonporous PTFE [186]
	• Cytoplast TXT-200 ® made of high-density PTFE [187]
	• Cytoplast Ti-250® made of Titanium-enforced high-density PTFE [147]
Advantages	• Excellent surgical handling
	• Effective clinical outcome
Disadvantages	• The need for the second surgery to remove the membrane
Natural bioresorbable membranes	
Examples	• BioGide® (Geistlich Biomaterials) made of bovine collagen [188]
	• BioMend® made of bovine Achilles tendon type I collagen [189]
	• Avitene ® made of microfibrillar hemostatic bovine corium type I collagen [190]
	• Collistat® made of bovine dermis collagen [191]
	• Oxidized cellulose membrane [163]
	• Dura mater membrane [164, 165]
	• Laminar bone membrane [166]
	• Chitosan-based membrane[161]
	• Gelatin-based membrane [162]
Advantages	• Low immunogenicity
	• Potential capability to augment the thickness of tissues
	• Hemostatic
	• Good interaction with different cell types including gingival fibroblast cells and periodontal ligament
Disadvantages	• Inadequate toughness
	• Low space maintenance
	• Relatively high cost
	• Difficult surgical handling
	• Possible rapid resorption
Synthetic bioresorbable membranes	
Examples	• GUIDOR® matrix barrier made of PDLLA and PLLA doped with acetyl tri-n-butyl citrate [181]
	• Resolut Adapt® made of PLGA [185]
	• Epi-Guide® made of PDLLA [192]
	• Vivosorb® made of Poly(D,L-lactide-ε-caprolactone) [193]
	• 3D printed PCL/PLGA/β-TCP membrane [183]
Advantages	• Controllable biodegradability
	• Low rigidity
	• Good surgical handling and manageability
	• Ability to encapsulate drugs
	• Processability
	• Possible good space maintainer
Disadvantages	• Unfavorable degradation product
	• Possible inflammatory response
	• Low biological activity

stability of Gore-Tex comes from this part, which acts as a barrier near the gingival flap [134].

The bilayer configuration of Gore-Tex membrane, with different porosity in each layer, established the principle of pore size anisotropy of a GTR membrane, which inspired the design of the GTR membrane developed in this thesis.

The outcomes of using ePTFE membranes were generally successful in different clinical studies [137–139]. Although some investigation indicated that the membrane could act similar to conventional flap surgery [140], there are other studies that exhibited its capability to considerably facilitate the periodontal regeneration after 3 months [141]. Some studies also reported observation of new cementum formation with inserting fibers [142]. There are also reports indicating other complications such as swelling and pain [143].

Integration of titanium reinforcements between the two layers of the Gore-Tex membrane has made it a better construct in terms of space maintenance and mechanical properties [142, 144]. Although some clinical studies reported that the incorporation of titanium reinforcements could not make much difference in the final clinical outcomes [137].

Other examples of non-resorbable membranes include TefGen-FD® made of dense nonporous PTFE [145], Cytoplast TXT-200® made of high-density PTFE [146], and Cytoplast Ti-250® made of titanium enforced high-density PTFE [147].

Since non-resorbable membranes do not degrade in the body, they have to be surgically removed in the follow-up phases of treatments. The second surgery can cause additional trauma in which the membrane removal interferes with the growth of new regenerated tissue [148]. Moreover, the second surgery is not favorable for patients in terms of additional cost, pain, and therapy period [148]. Thus, bioresorbable materials, made of synthetic or natural materials, became very popular.

7.7.2   Natural Bioresorbable Membrane

Collagen is one of the main natural materials in the production of resorbable GTR membranes due to its low immunogenicity, hemostatic, potential capability to augment the thickness of tissues, and the capability to interact with different cell types including gingival fibroblast cells and periodontal ligament [149, 150].

In a comparison study between collagen and PTFE, it was shown that while PTFE may inhibit the DNA synthesis of gingival fibroblasts and significantly decrease the synthesis of ECM, collagen could facilitate the gingival fibroblast proliferation [151]. The superior attachment capability of collagen membrane to osteoblasts has also been verified [152].

Animal studies showed that it took 8 weeks for BioGide® as a bovine collagen membrane to be degraded in the body, while this degradation time for a rat-tail collagen membrane was 4 weeks [153, 154]. However, both membranes facilitated periodontal regeneration. Although there were signs of chronic inflammation adjacent to the membranes, the signs were vanished after full degradation of the membranes [153, 154].

BioMend® as a bovine Achilles tendon type I collagen membrane with a pore size of 0.004 μm can also degrade in 4–8 weeks [155]. Good clinical results were reported regarding the use of this membrane [156–158]. However, since it does not have the capability of maintaining space, the clinical effectiveness of the membranes depends on the size of the defects [156–158].

Using Avitene® GTR membranes, made of microfibrillar hemostatic bovine corium type I collagen, some investigators reported the difficulty in surgical handling and also comparable effectivity of this membrane to the control group, evidenced by histological evaluations [159]. Similar results of non-effectivity were also reported using Collistat® hemostatic membrane made of bovine dermis collagen, in which rapid degradation (7 days) was another concern [160].

Among the non-collagen natural materials, chitosan showed good results to be used in resorbable membranes. For example, Kuo et al. examined chitosan membranes for GTR applications by preparing the membranes using thermally induced phase separation method and examined its function in the critical-sized skull defects of rats [161].

Combination of gelatin and HA was also used for GTR/GBR applications [162]. More specifically matrix of HA precipitate and gelatin was first lyophilized and dissolved in an organic solution and then went through the electrospinning procedure to make the membrane [162].

Many other natural materials have also been examined for use in GTR application. Some were not successful, such as oxidized cellulose [163], dura mater [164, 165], and laminar bone [166]. Our group has prepared the first 3D printed hydrogel membrane for GTR applications [167]. This membrane is specifically made to avoid penetration of keratinocytes into the connective tissue [167]. The material of the membrane is composed of gelatin, elastin, and sodium hyaluronate. The rationale behind selecting these three components is related to the properties of each material. Gelatin is low-cost and can act as a cell attractant for almost any cell type [168, 169]. Elastin has long-term stability, elasticity, and biological activity [170], while sodium hyaluronate facilitates chemical signaling among cells [171]. After performing the rheological analyses on many different compositions of these three components, a specific composition of 8% gelatin, 2% elastin, and 0.5% sodium hyaluronate (w/v) in aqueous solution has been selected as the most suitable and printable ink, which results in membranes with appropriate biocompatibility, flexibility, mechanical integrity, and secure surgical handling [167, 172].

The protocol and printing parameters of this ink are established in experimental approaches with the assist of rheological analyses of the composition [167, 172]. The printability of this composition allows the fabrication of membranes with various porosity, pore size, thickness, and configurations. However, a specific design is adopted for the GTR application, which is illustrated in Fig. 7.26 [167]. This six-layer membrane is printed with the strand distance of 0.6 μm for the first four layers and 0.9 μm for the last two layers. The strand angels are also different for different layers. More specifically, they are 45°, 135°, 0°, 90°, 0°, 90° from the first to the last layer. Considering the difference in the strand distance and printing angles for different layers, the first layers are composed of small pore size to accommodate the

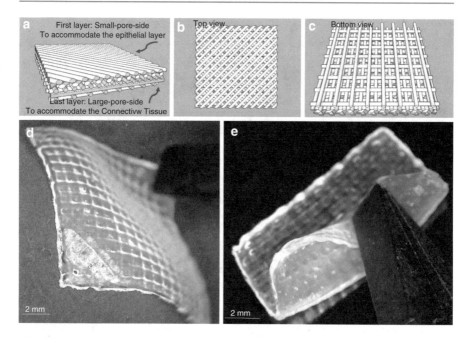

Fig. 7.26. (**a–c**): Design of the 3D printed GTR membrane. (**d–e**): Actual membrane made of 8% gelatin, 2% elastin, and 0.5% sodium hyaluronate (w/v) after printing. Reproduced from reference [167] with permission

epithelial layer and the last layers have a large pore size suitable for the growth of fibroblasts. The pores at the first layers are small enough to prevent the keratinocytes to drop into the connective tissue section. On the other hand, the pore size of the other side is appropriate for the penetration of fibroblasts to make the connective tissue [167].

As seen in Fig. 7.27, other than the pore size, the roughness values of the two sides of the membrane are different, which provide different environments for growing dissimilar types of cells. More specifically, the approximate value of the membrane pore size on one side is 140 μm, while this value for the other side is about 440 μm. The roughness value (R_a: arithmetical mean deviation of the roughness profile) for the side with small pores size is approximately 0.30 μm and is much higher (i.e., ~1.19 μm) for the side with a large pore size [167].

The thickness of this membrane is approximately 150 μm when measured by Electronic Digital Micrometer and confirmed using the 3D measuring laser microscopy. However, it should be noted that the thickness of the 3D printed membrane can be significantly higher at the edge, if the membrane includes a contour. Printing a contour (stacking the strands on top of each other at the edge of a construct) is a strategy to keep the integrity of such membrane, making it mechanically stronger. This is the case in this developed 3D printed GTR membrane.

More specifically, Fig. 7.28 presents the membrane's mechanical properties. Tensile modulus (elastic modulus) of 1.95 ± 0.55 MPa is measured for a

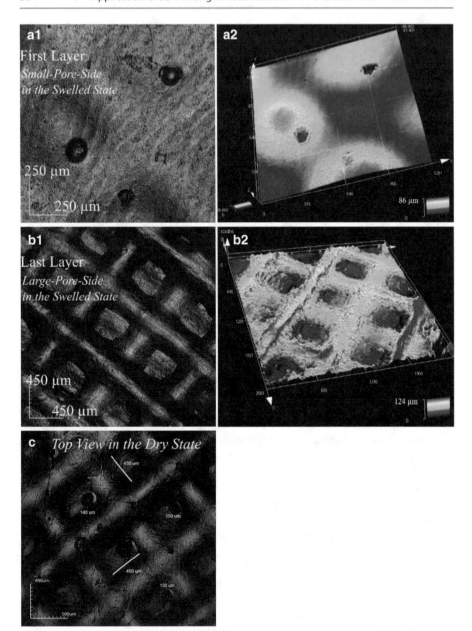

Fig. 7.27 Images of a 3D printed GTR membrane with different pore size of different sides taken by 3D measuring laser microscope. While in completely swelled state, the images of first and last layers demonstrate the pore size in two sides of the scaffold separately (**a1** and **b1**). The image taken in dry state (**c**) clearly shows the two different pore sizes of the membrane in one photo shot. Reproduced from reference [167] with permission

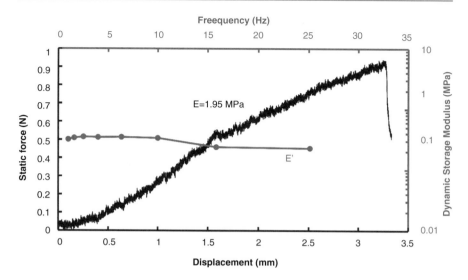

Fig. 7.28 Static and dynamic mechanical properties of a 3D printed GTR membrane, suggesting its proper surgical handing. Reproduced from reference [167] with permission

15 mm×15 mm membrane, with the maximum stress and maximum elongation at a breaking point of 1.15 ± 0.33 MPa and 60 ± 7% of the initial length, respectively. The linear (elastic) behavior of the membrane can be seen until it is torn. The dynamic storage modulus of the membrane using a DMA indicates that the storage modulus is not altered noticably by altering the frequency. The value of 314 ± 50 kPa is measured for the tensile storage modulus. Overall, the suitable mechanical property values indicate the proper surgical handling of this 3D printed construct, which is essential for its use as a GTR membrane.

A detailed in vitro analysis of the fabricated 3D printed construct confirms the GTR functionality of the membrane. More specifically, as seen in the histological analysis of Fig. 7.29b, keratinocytes can form a continuous epithelium layer on the side with smaller pore size and lower roughness. On the other hand, oral fibroblasts grow very well on the side with a large pore size and higher roughness, as demonstrated in Fig. 7.29c. In order to show the barrier function of a membrane, DAPI staining can be used as an identifier of any invasion and penetration of a cell from one side to the other. As shown in Fig. 7.29d, DAPI staining suggests that there is no sign of epithelial invasion into the inner layers of the membrane, which confirms the barrier functionality of the produced 3D printed construct as a GTR membrane.

7.7.3   Synthetic Bioresorbable Membranes

The disadvantages of natural membranes including their high cost, inadequate toughness, and incapability to act as a space maintenance [149], made investigators to continuously look for synthetic materials to make resorbable GTR/GBR membranes.

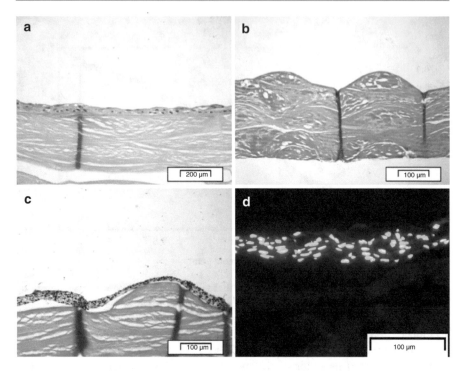

Fig. 7.29 Histological images of (**a**): 3D printed membrane before cell seeding. (**b**): Epithelium layers on the side with small pore size. (**c**): Fibroblast growth on the side with large pore size ((**a–c**): Hematoxylin and eosin staining). (**d**): DAPI staining of the cell nuclei on top of the 3D printed membrane confirmed that cells did not penetrate from one side to the other, indicating the barrier functionality of the membrane to be used for GTR applications. Reproduced from reference [167] with permission

Aliphatic polyesters are the main group of materials used for fabrication of synthetic resorbable membranes [131]. More specifically, PCL, PLA, poly(hydroxyl butyric acid), PGA, poly(hydroxyl valeric acid), and their copolymers play the important role in such membranes [131].

The degradation products of these materials can cause unfavorable inflammatory response of living tissues [131, 173]. There are reports that showed using PLA and PGA membranes can cause a reduced defect fill compared to ePTFE membranes [174, 175].

The membranes made by synthetic materials are not as biologically active as the ones made by natural materials [176]. However, they have advantages such as controllable biodegradability, low rigidity, good surgical handling and manageability, and the ability to encapsulate drugs and processability [174, 177–179].

Guidor® Matrix Barrier, a bi-layered membrane made by the combination of poly(D,L-lactide) (PDLLA) and PLLA which is doped with acetyl tri-n-butyl citrate, is known as the first commercially available synthetic GTR membranes [180, 181].

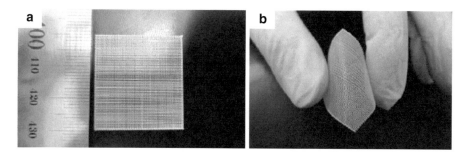

Fig. 7.30 (**a**): 3D printed PCL/PLGA/β-TCP GBR membrane. (**b**): A demonstration of the membrane flexibility. Reproduced from reference [182] with permission

The degradation time of this membrane is 13 months, and it can hold its barrier characteristics for approximately 6 weeks [180].

Shim et al., fabricated a 3D printed GBR membrane using a combination of PCL, PLGA, and β-TCP as shown in Fig. 7.30 [182]. This membrane was loaded by rhBMP-2 and used to reconstruct calvaria defects in rabbits. The result showed that the membrane has good space-making ability and could heal the calvaria defects within 8 weeks [182].

The same research group evaluated this 3D printed PCL/PLGA/β-TCP membrane in a beagle implant model and compared its function with the function of a collagen membrane [183]. They concluded that this 3D printed membrane can maintain its reliability and structure better than the collagen membrane and thus more appropriate for GBR applications [183].

Our group has combined the 3D printed PCL membrane with hydrogels as a construct that can be used for interfacial tissue engineering. For example, in part of a recent patent application entitled "Medical and dental integrated multiphasic biomaterials for single or multi-tissue reconstruction/regeneration," we have developed an integrated biphasic construct, as shown in Fig. 7.15, which can be used as a GBR membrane (other than the specific application of PDL/bone regeneration explained in Sect. 7.5.1) [31]. Moreover, our group has recently fabricated a 3D printed PCL/gelatin membrane in which a layer of gelatin is attached to one side of a PCL mesh (Fig. 7.31). In vitro and in vivo analyses confirmed the function of the construct as a GBR membrane [184]. More specifically, in vitro experiments showed that the cells cultured on one side could not penetrate to the other side, indicating the barrier functionality of the membrane [184]. Moreover, the animal study at the bone defect of the dog tibia showed that using this membrane could increase the mean bone density by three times [184]. The method of adhesion between the gelatin layer and 3D printed PCL mesh was similar to what was explained in Sect. 7.3.1 for attaching the hard scaffold to the soft one in the osteomucosal model.

Some other examples of synthetic GTR/GBR membranes are listed in Table 7.1.

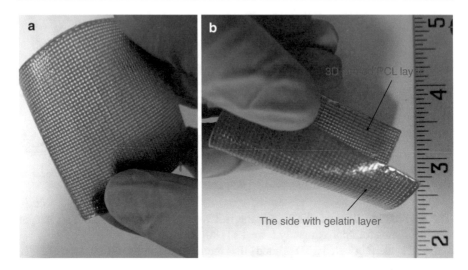

Fig. 7.31   3D printed PCL/gelatin GBR membrane

7.8   Summary

Interfacial structures are important constituents of oral and maxillofacial tissues. Several regenerative strategies have been developed to fabricate osteomucosal, osteochondral, and periodontal complexes. Among the most successful strategies are tissue engineering, the organ germ method, and cell sheet engineering. These strategies provide appropriate combinations of cells, signals, and scaffolds to create the desired constructs with optimal structure and function. 3D printing can provide a unique approach for making constructs with different properties in different regions to be able to accommodate the different cells required for the reconstruction of multi- and interfacial tissues in the oral and maxillofacial region.

This chapter presented a brief introduction to the regeneration of a heterogenous interface using the 3D printing method. Then, osteomucosal, osteochondral, periodontal, and whole tooth regeneration, as important examples of multi- and interfacial tissues in oral and maxillofacial regions, were discussed, and the relevant strategies to reconstruct them, including the 3D printing approach, were elaborated. The final section of this chapter was dedicated to reviewing GBR/GTR membranes as an effective tool to regenerate interfacial tissue and the applications of 3D printing in their production.

Despite the promising advances in the regeneration of oral and maxillofacial multi- and interfacial tissues, clinical applications of these constructs require further investigation.

References

1. Phillips JE, Burns KL, Le Doux JM, Guldberg RE, García AJ. Engineering graded tissue interfaces. Proc Natl Acad Sci. 2008;105(34):12170–5.
2. Vaquette C, Fan W, Xiao Y, Hamlet S, Hutmacher DW, Ivanovski S. A biphasic scaffold design combined with cell sheet technology for simultaneous regeneration of alveolar bone/ periodontal ligament complex. Biomaterials. 2012;33(22):5560–73.
3. Requicha JF, Viegas CA, Hede S, Leonor IB, Reis RL, Gomes ME. Design and characterization of a biodegradable double-layer scaffold aimed at periodontal tissue-engineering applications. J Tissue Eng Regen Med. 2016;10(5):392–403.
4. Seidi A, Ramalingam M, Elloumi-Hannachi I, Ostrovidov S, Khademhosseini A. Gradient biomaterials for soft-to-hard interface tissue engineering. Acta Biomater. 2011;7(4):1441–51.
5. Gibble J, Ness P. Fibrin glue: the perfect operative sealant? Transfusion. 1990;30(8):741–7.
6. Spotnitz WD. Fibrin sealant: past, present, and future: a brief review. World J Surg. 2010;34(4):632–4.
7. Lauto A, Mawad D, Foster LJR. Adhesive biomaterials for tissue reconstruction. J Chem Technol Biotechnol Int Res Process Environ Clean Technol. 2008;83(4):464–72.
8. Lu HH, Subramony SD, Boushell MK, Zhang X. Tissue engineering strategies for the regeneration of orthopedic interfaces. Ann Biomed Eng. 2010;38(6):2142–54.
9. Thal R. Process for attaching tissue to bone using a captured-loop knotless suture anchor assembly. Google Patents. 1999.
10. Panseri S, Russo A, Cunha C, Bondi A, Di Martino A, Patella S, Kon E. Osteochondral tissue engineering approaches for articular cartilage and subchondral bone regeneration. Knee Surg Sports Traumatol Arthrosc. 2012;20(6):1182–91.
11. Eke PI, Dye BA, Wei L, Slade GD, Thornton-Evans GO, Borgnakke WS, Taylor GW, Page RC, Beck JD, Genco RJ. Update on prevalence of periodontitis in adults in the United States: NHANES 2009 to 2012. J Periodontol. 2015;86(5):611–22.
12. Boffano P, Roccia F, Zavattero E, Dediol E, Uglešić V, Kovačič Ž, Vesnaver A, Konstantinović VS, Petrović M, Stephens J. European Maxillofacial Trauma (EURMAT) project: a multicentre and prospective study. J Cranio-Maxillofac Surg. 2015;43(1):62–70.
13. Jemal A, Siegel R, Xu J, Ward E. Cancer statistics, 2010. CA Cancer J Clin. 2010;60(5):277–300.
14. Parkin DM. Global cancer statistics in the year 2000. Lancet Oncol. 2001;2(9):533–43.
15. Guerra MFM, Gías LN, Campo FRG, Pérez JS. Marginal and segmental mandibulectomy in patients with oral cancer: a statistical analysis of 106 cases. J Oral Maxillofac Surg. 2003;61(11):1289–96.
16. Bottino MC, Thomas V. Membranes for periodontal regeneration--a materials perspective. Front Oral Biol. 2015;17:90–100.
17. Needleman IG, Worthington HV, Giedrys-Leeper E, Tucker RJ. Guided tissue regeneration for periodontal infra-bony defects. Cochrane Database Syst Rev. 2006;2:CD001724.
18. Rispoli L, Fontana F, Beretta M, Poggio CE, Maiorana C. Surgery guidelines for barrier membranes in guided bone regeneration (GBR). J Otolaryngol Rhinol. 2015;1:008.
19. Tayebi L, Rasoulianboroujeni M, Cui Z, Ye H. 3D-printed thick structured gelatin membrane for engineering of heterogeneous tissues. Mater Lett. 2018;217:39–43.
20. Almela T, Brook IM, Moharamzadeh K. Development of three-dimensional tissue engineered bone-oral mucosal composite models. J Mater Sci Mater Med. 2016;27(4):65.
21. Moharamzadeh K. 14 - Oral mucosa tissue engineering. In: Tayebi L, Moharamzadeh K, editors. Biomaterials for Oral and dental tissue engineering. Woodhead Publishing; 2017. p. 223–44.
22. Bae S, Sun S, Aghaloo T, Oh J-E, McKenna CE, Kang MK, Shin K-H, Tetradis S, Park N-H, Kim RH. Development of oral osteomucosal tissue constructs in vitro and localization of fluorescently-labeled bisphosphonates to hard and soft tissue. Int J Mol Med. 2014;34(2):559–63.

23. Peng HT, Shek PN. Novel wound sealants: biomaterials and applications. Expert Rev Med Devices. 2010;7(5):639–59.
24. Bouten PJ, Zonjee M, Bender J, Yauw ST, van Goor H, van Hest JC, Hoogenboom R. The chemistry of tissue adhesive materials. Prog Polym Sci. 2014;39(7):1375–405.
25. Farrar DF. Bone adhesives for trauma surgery: a review of challenges and developments. Int J Adhes Adhes. 2012;33:89–97.
26. Spotnitz WD, Welker R. Clinical uses of fibrin sealant, transfusion therapy: clinical principles and practice (Mintz PD, Ed), AABB. Bethesda; 1999. p. 199–222.
27. Montanaro L, Arciola C, Cenni E, Ciapetti G, Savioli F, Filippini F, Barsanti L. Cytotoxicity, blood compatibility and antimicrobial activity of two cyanoacrylate glues for surgical use. Biomaterials. 2000;22(1):59–66.
28. Oliveira CLD, Santos CHMD, Bezerra FMM, Bezerra MM, Rodrigues LDL. Utilization of cyanoacrylates adhesives in skin suture. Revista Brasileira de Cirurgia Plástica. 2010;25(3):573–6.
29. Ferreira P, Pereira R, Coelho J, Silva AF, Gil M. Modification of the biopolymer castor oil with free isocyanate groups to be applied as bioadhesive. Int J Biol Macromol. 2007;40(2):144–52.
30. Bergmann ED, Ginsburg D, Pappo R. The Michael reaction. Org React. 1959;10:179.
31. Tayebi L, Dashtimoghadam E, Fahimipour F. WO/2018/209101, https://patentscope.wipo.int/search/en/detail.jsf?docId=WO2018209101, Medical and dental integrated multiphasic biomaterials for single or multi-tissue reconstruction/regeneration. 2018.
32. Moharamzadeh K, Brook I, Van Noort R, Scutt A, Thornhill M. Tissue-engineered oral mucosa: a review of the scientific literature. J Dent Res. 2007;86(2):115–24.
33. Moharamzadeh K, Colley H, Murdoch C, Hearnden V, Chai W, Brook I, Thornhill M, Macneil S. Tissue-engineered oral mucosa. J Dent Res. 2012;91(7):642–50.
34. Whang K, Healy K, Elenz D, Nam E, Tsai D, Thomas C, Nuber G, Glorieux F, Travers R, Sprague S. Engineering bone regeneration with bioabsorbable scaffolds with novel microarchitecture. Tissue Eng. 1999;5(1):35–51.
35. Yang S, Leong K-F, Du Z, Chua C-K. The design of scaffolds for use in tissue engineering. Part I. traditional factors. Tissue Eng. 2001;7(6):679–89.
36. Zeltinger J, Sherwood JK, Graham DA, Müeller R, Griffith LG. Effect of pore size and void fraction on cellular adhesion, proliferation, and matrix deposition. Tissue Eng. 2001;7(5):557–72.
37. Van Tienen TG, Heijkants RG, Buma P, de Groot JH, Pennings AJ, Veth RP. Tissue ingrowth and degradation of two biodegradable porous polymers with different porosities and pore sizes. Biomaterials. 2002;23(8):1731–8.
38. Longley R, Ferreira A, Gentile P. Recent approaches to the manufacturing of biomimetic multi-phasic scaffolds for osteochondral regeneration. Int J Mol Sci. 2018;19(6):1755.
39. Benders KE, van Weeren PR, Badylak SF, Saris DB, Dhert WJ, Malda J. Extracellular matrix scaffolds for cartilage and bone regeneration. Trends Biotechnol. 2013;31(3):169–76.
40. Gillette BM, Rossen NS, Das N, Leong D, Wang M, Dugar A, Sia SK. Engineering extracellular matrix structure in 3D multiphase tissues. Biomaterials. 2011;32(32):8067–76.
41. Kim B-S, Park I-K, Hoshiba T, Jiang H-L, Choi Y-J, Akaike T, Cho C-S. Design of artificial extracellular matrices for tissue engineering. Prog Polym Sci. 2011;36(2):238–68.
42. Fuentes-Mera L, Camacho A, Engel E, Pérez-Silos V, Lara-Arias J, Marino-Martínez I, Peña-Martínez V. Therapeutic potential of articular cartilage regeneration using tissue engineering based on multiphase designs. In: Cartilage tissue engineering and regeneration techniques. IntechOpen; 2019.
43. Castro NJ, Hacking SA, Zhang LG. Recent progress in interfacial tissue engineering approaches for osteochondral defects. Ann Biomed Eng. 2012;40(8):1628–40.
44. Swieszkowski W, Tuan BHS, Kurzydlowski KJ, Hutmacher DW. Repair and regeneration of osteochondral defects in the articular joints. Biomol Eng. 2007;24(5):489–95.
45. Yousefi AM, Hoque ME, Prasad RG, Uth N. Current strategies in multiphasic scaffold design for osteochondral tissue engineering: a review. J Biomed Mater Res A. 2015;103(7):2460–81.

46. Grayson WL, Bhumiratana S, Chao PG, Hung CT, Vunjak-Novakovic G. Spatial regulation of human mesenchymal stem cell differentiation in engineered osteochondral constructs: effects of pre-differentiation, soluble factors and medium perfusion. Osteoarthr Cartil. 2010;18(5):714–23.

47. Lee M, Wu BM. Recent advances in 3D printing of tissue engineering scaffolds. In: Computer-aided tissue engineering. Springer; 2012. p. 257–67.

48. Mauck RL, Soltz MA, Wang CC, Wong DD, Chao P-HG, Valhmu WB, Hung CT, Ateshian GA. Functional tissue engineering of articular cartilage through dynamic loading of chondrocyte-seeded agarose gels. J Biomech Eng. 2000;122(3):252–60.

49. Harley BA, Lynn AK, Wissner-Gross Z, Bonfield W, Yannas IV, Gibson LJ. Design of a multiphase osteochondral scaffold III: fabrication of layered scaffolds with continuous interfaces. J Biomed Mater Res Part A. 2010;92(3):1078–93.

50. Harley BA, Lynn AK, Wissner-Gross Z, Bonfield W, Yannas IV, Gibson LJ. Design of a multiphase osteochondral scaffold. II. Fabrication of a mineralized collagen–glycosaminoglycan scaffold. J Biomed Mater Res A. 2010;92(3):1066–77.

51. Wang Y, Meng H, Yuan X, Peng J, Guo Q, Lu S, Wang A. Fabrication and in vitro evaluation of an articular cartilage extracellular matrix-hydroxyapatite bilayered scaffold with low permeability for interface tissue engineering. Biomed Eng Online. 2014;13(1):80.

52. Miyagi S, Takagi M, Tensho K, Wakitani S. Construction of an osteochondral-like tissue graft combining β-tricalcium phosphate block and scaffold-free mesenchymal stem cell sheet. J Orthop Sci. 2013;18(3):471–7.

53. Niyama K, Ide N, Onoue T, Okabe T, Wakitani S, Takagi M. Construction of osteochondral-like tissue graft combining β-tricalcium phosphate block and scaffold-free centrifuged chondrocyte cell sheet. J Orthop Sci. 2011;16(5):613–21.

54. HO SAEY TUAN (2009-06-28). Tissue engineering of an osteochondral transplant by using a cell/scaffold construct. ScholarBank@NUS Repository. 2009.

55. Jeon JE, Vaquette C, Theodoropoulos C, Klein TJ, Hutmacher DW. Multiphasic construct studied in an ectopic osteochondral defect model. J R Soc Interface. 2014;11(95):20140184.

56. Liu X-g, Jiang H-k. Preparation of an osteochondral composite with mesenchymal stem cells as the single-cell source in a double-chamber bioreactor. Biotechnol Lett. 2013;35(10):1645–53.

57. Shim J-H, Lee J-S, Kim JY, Cho D-W. Bioprinting of a mechanically enhanced three-dimensional dual cell-laden construct for osteochondral tissue engineering using a multi-head tissue/organ building system. J Micromech Microeng. 2012;22(8):085014.

58. Nguyen LH, Kudva AK, Saxena NS, Roy K. Engineering articular cartilage with spatially-varying matrix composition and mechanical properties from a single stem cell population using a multi-layered hydrogel. Biomaterials. 2011;32(29):6946–52.

59. Chen J, Chen H, Li P, Diao H, Zhu S, Dong L, Wang R, Guo T, Zhao J, Zhang J. Simultaneous regeneration of articular cartilage and subchondral bone in vivo using MSCs induced by a spatially controlled gene delivery system in bilayered integrated scaffolds. Biomaterials. 2011;32(21):4793–805.

60. Levy YD, Görtz S, Pulido PA, McCauley JC, Bugbee WD. Do fresh osteochondral allografts successfully treat femoral condyle lesions? Clin Orthop Relat Res. 2013;471(1):231–7.

61. Fuchs DJ. Osteochondral allograft transplantation in the ankle: a review of current practice. 2015.

62. Saltzman BM, Riboh JC, Cole BJ, Yanke AB. Humeral head reconstruction with osteochondral allograft transplantation. Arthroscopy J Arthroscopic Relat Surg. 2015;31(9):1827–34.

63. Puskas GJ, Giles JW, Degen RM, Johnson JA, Athwal GS. Humeral head reconstruction for hill-Sachs defects: a biomechanical comparison of 2 fixation techniques for bone grafting. Arthroscopy J Arthroscopic Relat Surg. 2014;30(1):22–8.

64. Kim J-B, Park J-S, Hong C-H, Kwon S-W, Soh J-W, Nho J-H, Lee C-J. Osteochondral lesion of humeral head associated with shoulder internal impingement: report of three cases. Kor J Sports Med. 2014;32(1):59–64.

65. Doral MN, Bilge O, Verdonk P, Dönmez G, Batmaz G. Osteochondral talus defects: treatment by biodegradable scaffolds, sports injuries: prevention, diagnosis. Treat Rehabil. 2015:1773–8.
66. Weigelt L, Siebenlist S, Hensler D, Imhoff A, Vogt S. Treatment of osteochondral lesions in the elbow: results after autologous osteochondral transplantation. Arch Orthop Trauma Surg. 2015;135(5):627–34.
67. Bancroft LW, Pettis C, Wasyliw C, Varich L. Osteochondral lesions of the elbow. In: Seminars in musculoskeletal radiology; 2013. p. 446–54.
68. Plath JE, Vogt S. Osteochondral lesion. Surgical Atlas of Sports Orthopaedics and Sports Traumatology: Springer; 2015. p. 119–22.
69. Zlotolow DA, Bae DS. Osteochondral autograft transplantation in the elbow. J Hand Surg Am. 2014;39(2):368.
70. Carey JL, Grimm NL. Treatment algorithm for osteochondritis dissecans of the knee. Clin Sports Med. 2014;33(2):375–82.
71. Dormer NH, Busaidy K, Berkland CJ, Detamore MS. Osteochondral interface regeneration of rabbit mandibular condyle with bioactive signal gradients. J Oral Maxillofac Surg. 2011;69(6):e50–7.
72. Hollister SJ, Levy RA, Chu TM, Halloran JW, Feinberg SE. An image-based approach for designing and manufacturing craniofacial scaffolds. Int J Oral Maxillofac Surg. 2000;29(1):67–71.
73. Schek RM, Taboas JM, Segvich SJ, Hollister SJ, Krebsbach PH. Engineered osteochondral grafts using biphasic composite solid free-form fabricated scaffolds. Tissue Eng. 2004;10(9–10):1376–85.
74. Schek R, Taboas J, Hollister SJ, Krebsbach P. Tissue engineering osteochondral implants for temporomandibular joint repair. Orthod Craniofac Res. 2005;8(4):313–9.
75. Williams JM, Adewunmi A, Schek RM, Flanagan CL, Krebsbach PH, Feinberg SE, Hollister SJ, Das S. Bone tissue engineering using polycaprolactone scaffolds fabricated via selective laser sintering. Biomaterials. 2005;26(23):4817–27.
76. Hollister SJ, Lin C, Saito E, Lin C, Schek R, Taboas J, Williams J, Partee B, Flanagan C, Diggs A. Engineering craniofacial scaffolds. Orthod Craniofac Res. 2005;8(3):162–73.
77. Smith M, Flanagan C, Kemppainen J, Sack J, Chung H, Das S, Hollister S, Feinberg S. Computed tomography-based tissue-engineered scaffolds in craniomaxillofacial surgery. Int J Med Robot Comput Assist Surg. 2007;3(3):207–16.
78. Das S, Adewunmi A, Williams J, Flanagan C, Engel A, Hollister S. Mechanical and structural properties of polycaprolactone scaffolds made by selective laser sintering. In: Proceedings of the 7th world biomaterials congress, Sydney; 2004.
79. Nanci A, Bosshardt DD. Structure of periodontal tissues in health and disease. Periodontol. 2006;40(1):11–28.
80. Eke P, Zhang X, Lu H, Wei L, Thornton-Evans G, Greenlund K, Holt J, Croft J. Predicting periodontitis at state and local levels in the United States. J Dent Res. 2016;95(5):515–22.
81. Vaquette C, Pilipchuk SP, Bartold PM, Hutmacher DW, Giannobile WV, Ivanovski S. Tissue engineered constructs for periodontal regeneration: current status and future perspectives. Adv Healthc Mater. 2018;7(21):1800457.
82. Mombelli A. Periodontitis as an infectious disease: specific features and their implications. Oral Dis. 2003;9:6–10.
83. Kuo L-C, Polson AM, Kang T. Associations between periodontal diseases and systemic diseases: a review of the inter-relationships and interactions with diabetes, respiratory diseases, cardiovascular diseases and osteoporosis. Public Health. 2008;122(4):417–33.
84. Wang H-L, Greenwell H, Fiorellini J, Giannobile W, Offenbacher S, Salkin L, Townsend C, Sheridan P, Genco RJ, Research S. Periodontal regeneration. J Periodontol. 2005;76(9):1601.
85. American Academy of Periodontology. Comprehensive periodontal therapy: a statement by the American Academy of periodontology. J Periodontol. 2011;82(7):943–9.
86. Nyman S. Bone regeneration using the principle of guided tissue regeneration. J Clin Periodontol. 1991;18(6):494–8.

87. Reynolds MA, Kao RT, Camargo PM, Caton JG, Clem DS, Fiorellini JP, Geisinger ML, Mills MP, Nares S, Nevins ML. Periodontal regeneration–intrabony defects: a consensus report from the AAP regeneration workshop. J Periodontol. 2015;86:S105–7.

88. Needleman I, Worthington HV, Giedrys-Leeper E, Tucker R. Guided tissue regeneration for periodontal infra-bony defects. In: Cochrane database of systematic reviews (2); 2006.

89. Cortellini P, Carnevale G, Sanz M, Tonetti MS. Treatment of deep and shallow intrabony defects a multicenter randomized controlled clinical trial. J Clin Periodontol. 1998;25(12):981–7.

90. Trombelli L. Which reconstructive procedures are effective for treating the periodontal intraosseous defect? Periodontol. 2005;37(1):88–105.

91. Murphy KG, Gunsolley JC. Guided tissue regeneration for the treatment of periodontal intrabony and furcation defects. A systematic review. Ann Periodontol. 2003;8(1):266–302.

92. Corbella S, Weinstein R, Francetti L, Taschieri S, Del Fabbro M. Periodontal regeneration in aggressive periodontitis patients: a systematic review of the literature. J Investig Clin Dent. 2017;8(4):e12245.

93. Langer R, Vacanti JP. Tissue engineering. Science. 1993;260(5110):920–6.

94. Babo PS, Reis RL, Gomes ME. Periodontal tissue engineering: current strategies and the role of platelet rich hemoderivatives. J Mater Chem B. 2017;5(20):3617–28.

95. Goudouri O-M, Kontonasaki E, Boccaccini AR. Layered scaffolds for periodontal regeneration. In: Tayebi L, Moharamzadeh K, editors. Biomaterials for oral and dental tissue engineering. Woodhead Publishing; 2017. p. 279–95.

96. Liu J, Ruan J, Weir MD, Ren K, Schneider A, Wang P, Oates TW, Chang X, Xu HHK. Periodontal bone-ligament-cementum regeneration via scaffolds and stem cells. Cell. 2019;8(6):537.

97. Liu Z, Yin X, Ye Q, He W, Ge M, Zhou X, Hu J, Zou S. Periodontal regeneration with stem cells-seeded collagen-hydroxyapatite scaffold. J Biomater Appl. 2016;31(1):121–31.

98. Reis ECC, Borges AP, Araújo MV, Mendes VC, Guan L, Davies JE. Periodontal regeneration using a bilayered PLGA/calcium phosphate construct. Biomaterials. 2011;32(35):9244–53.

99. Zhang Y, Miron RJ, Li S, Shi B, Sculean A, Cheng X. Novel Meso porous BioGlass/silk scaffold containing ad PDGF-B and ad BMP 7 for the repair of periodontal defects in beagle dogs. J Clin Periodontol. 2015;42(3):262–71.

100. Wu M, Wang J, Zhang Y, Liu H, Dong F. Mineralization induction of gingival fibroblasts and construction of a sandwich tissue-engineered complex for repairing periodontal defects. Med Sci Monit Int Med J Exp Clin Res. 2018;24:1112.

101. Yang J, Yamato M, Shimizu T, Sekine H, Ohashi K, Kanzaki M, Ohki T, Nishida K, Okano T. Reconstruction of functional tissues with cell sheet engineering. Biomaterials. 2007;28(34):5033–43.

102. Chen G, Qi Y, Niu L, Di T, Zhong J, Fang T, Yan W. Application of the cell sheet technique in tissue engineering. Biomed Rep. 2015;3(6):749–57.

103. Okano T, Yamada N, Sakai H, Sakurai Y. A novel recovery system for cultured cells using plasma-treatedpolystyrene dishes grafted with poly(N-isopropylacrylamide). J Biomed Mater Res. 1993;27(10):1243–51.

104. Sawa Y, Miyagawa S. Present and future perspectives on cell sheet-based myocardial regeneration therapy. BioMed Research International; 2013.

105. Zavala J, Jaime GL, Barrientos CR, Valdez-Garcia J. Corneal endothelium: developmental strategies for regeneration. Eye. 2013;27(5):579–88.

106. Wang J, Zhang R, Shen Y, Xu C, Qi S, Lu L, Wang R, Xu Y. Recent advances in cell sheet technology for periodontal regeneration. Curr Stem Cell Res Ther. 2014;9(3):162–73.

107. Zhang H, Liu S, Zhu B, Xu Q, Ding Y, Jin Y. Composite cell sheet for periodontal regeneration: crosstalk between different types of MSCs in cell sheet facilitates complex periodontal-like tissue regeneration. Stem Cell Res Ther. 2016;7(1):168.

108. Iwata T, Yamato M, Washio K, Yoshida T, Tsumanuma Y, Yamada A, Onizuka S, Izumi Y, Ando T, Okano T. Periodontal regeneration with autologous periodontal ligament-derived cell sheets–a safety and efficacy study in ten patients. Regen Therapy. 2018;9:38–44.

109. Raju R, Oshima M, Inoue M, Morita T, Huijiao Y, Waskitho A, Baba O, Inoue M, Matsuka Y. Three-dimensional periodontal tissue regeneration using a bone-ligament complex cell sheet. Sci Rep. 2020;10(1):1–16.
110. Amirabad LM, Zarrintaj P, Lindemuth A, Tayebi L. Whole tooth engineering. Springer, Cham: Appl Biomed Eng Dent; 2020. p. 443–62.
111. Chen G, Chen J, Yang B, Li L, Luo X, Zhang X, Feng L, Jiang Z, Yu M, Guo W. Combination of aligned PLGA/gelatin electrospun sheets, native dental pulp extracellular matrix and treated dentin matrix as substrates for tooth root regeneration. Biomaterials. 2015;52:56–70.
112. Kuchler-Bopp S, Larrea A, Petry L, Idoux-Gillet Y, Sebastian V, Ferrandon A, Schwinté P, Arruebo M, Benkirane-Jessel N. Promoting bioengineered tooth innervation using nanostructured and hybrid scaffolds. Acta Biomater. 2017;50:493–501.
113. Kitasako Y, Shibata S, Cox C, Tagami J. Location, arrangement and possible function of interodontoblastic collagen fibres in association with calcium hydroxide-induced hard tissue bridges. Int Endod J. 2002;35(12):996–1004.
114. Demarco FF, Conde MCM, Cavalcanti BN, Casagrande L, Sakai VT, Nör JE. Dental pulp tissue engineering. Braz Dent J. 2011;22(1):3–13.
115. Kim NR, Lee DH, Chung P-H, Yang H-C. Distinct differentiation properties of human dental pulp cells on collagen, gelatin, and chitosan scaffolds. Oral Surg Oral Med Oral Pathol Oral Radiol Endod. 2009;108(5):e94–e100.
116. Chen H, Fan M. Chitosan/carboxymethyl cellulose polyelectrolyte complex scaffolds for pulp cells regeneration. J Bioact Compat Polym. 2007;22(5):475–91.
117. Ravindran S, Song Y, George A. Development of three-dimensional biomimetic scaffold to study epithelial–mesenchymal interactions. Tissue Eng A. 2010;16(1):327–42.
118. Hu B, Nadiri A, Kuchler-Bopp S, Perrin-Schmitt F, Peters H, Lesot H. Tissue engineering of tooth crown, root, and periodontium. Tissue Eng. 2006;12(8):2069–75.
119. Ikeda E, Tsuji T. Growing bioengineered teeth from single cells: potential for dental regenerative medicine. Expert Opin Biol Ther. 2008;8(6):735–44.
120. Hu L, Liu Y, Wang S. Stem cell-based tooth and periodontal regeneration. Oral Dis. 2018;24(5):696–705.
121. Yang B, Chen G, Li J, Zou Q, Xie D, Chen Y, Wang H, Zheng X, Long J, Tang W. Tooth root regeneration using dental follicle cell sheets in combination with a dentin matrix-based scaffold. Biomaterials. 2012;33(8):2449–61.
122. Wei F, Song T, Ding G, Xu J, Liu Y, Liu D, Fan Z, Zhang C, Shi S, Wang S. Functional tooth restoration by allogeneic mesenchymal stem cell-based bio-root regeneration in swine. Stem Cells Dev. 2013;22(12):1752–62.
123. Meng H, Hu L, Zhou Y, Ge Z, Wang H, Wu C-t, Jin J. A Sandwich structure of human dental pulp stem cell sheet, treated dentin matrix, and matrigel for tooth root regeneration. Stem Cells Dev. 2020;29(8):521–32.
124. Hurley LA, Stinchfield FE, Bassett CAL, Lyon WH. The role of soft tissues in osteogenesis: an experimental study of canine spine fusions. JBJS. 1959;41(7):1243–66.
125. Gottlow J, Nyman S, Karring T, Lindhe J. New attachment formation as the result of controlled tissue regeneration. J Clin Periodontol. 1984;11(8):494–503.
126. Villar CC, Cochran DL. Regeneration of periodontal tissues: guided tissue regeneration. Dent Clin. 2010;54(1):73–92.
127. Retzepi M, Donos N. Guided bone regeneration: biological principle and therapeutic applications. Clin Oral Implants Res. 2010;21(6):567–76.
128. Buser D, Dahlin C, Schenk R. Guided bone regeneration. Chicago Quintessence; 1994.
129. Bottino MC, Thomas V, Schmidt G, Vohra YK, Chu T-MG, Kowolik MJ, Janowski GM. Recent advances in the development of GTR/GBR membranes for periodontal regeneration—a materials perspective. Dent Mater. 2012;28(7):703–21.
130. Caballé-Serrano J, Munar-Frau A, Ortiz-Puigpelat O, Soto-Penaloza D, Peñarrocha M, Hernández-Alfaro F. On the search of the ideal barrier membrane for guided bone regeneration. J Clin Exp Dent. 2018;10(5):e477.

131. Wang J, Wang L, Zhou Z, Lai H, Xu P, Liao L, Wei J. Biodegradable polymer membranes applied in guided bone/tissue regeneration: a review. Polymers. 2016;8(4):115.
132. Bartee CM, Bartee BK. PTFE composite multi-layer material. Google Patents. 2010.
133. Williams DF, Roaf R, Maisels D, Temple LJ, Wright J. Implants in surgery. London: Saunders; 1973.
134. Aurer A, Jorgić-Srdjak K. Membranes for periodontal regeneration. Acta Stomatol Croat. 2005;39(1):107–12.
135. Schliephake H, Dard M, Planck H, Hierlemann H, Jakob A. Guided bone regeneration around endosseous implants using a resorbable membrane vs a PTFE membrane. Clin Oral Implants Res. 2000;11(3):230–41.
136. Scantlebury TV. 1982-1992: a decade of technology development for guided tissue regeneration. J Periodontol. 1993;64:1129–37.
137. Becker W, Becker BE. Periodontal regeneration: A contemporary re-evaluation. Periodontol. 1999;19(1):104–14.
138. Buser D, Dula K, Hess D, Hirt HP, Belser UC. Localized ridge augmentation with autografts and barrier membranes. Periodontol. 1999;19(1):151–63.
139. Trombelli L. Periodontal regeneration in gingival recession defects. Periodontology. 1999;2000(19):138–50.
140. Ratka-Krüger P, Neukranz E, Raetzke P. Guided tissue regeneration procedure with bioresorbable membranes versus conventional flap surgery in the treatment of infrabony periodontal defects. J Clin Periodontol. 2000;27(2):120–7.
141. Gottlow J, Nyman S, Lindhe J, Karring T, Wennström J. New attachment formation in the human periodontium by guided tissue regeneration case reports. J Clin Periodontol. 1986;13(6):604–16.
142. Cortellini P, Prato GP, Tonetti MS. Periodontal regeneration of human intrabony defects with titanium reinforced membranes. A controlled clinical trial. J Periodontol. 1995;66(9):797–803.
143. Murphy KG. Postoperative healing complications associated with Gore-Tex periodontal material. Part I. incidence and characterization. Int J Periodont Restor Dent. 1995;15(4):363–75.
144. Hardwick R, Hayes BK, Flynn C. Devices for dentoalveolar regeneration: an up-to-date literature review. J Periodontol. 1995;66(6):495–505.
145. Rakhmatia YD, Ayukawa Y, Furuhashi A, Koyano K. Current barrier membranes: titanium mesh and other membranes for guided bone regeneration in dental applications. J Prosthodont Res. 2013;57(1):3–14.
146. Barber HD, Lignelli J, Smith BM, Bartee BK. Using a dense PTFE membrane without primary closure to achieve bone and tissue regeneration. J Oral Maxillofac Surg. 2007;65(4):748–52.
147. Sheikh Z, Khan AS, Roohpour N, Glogauer M, Rehman IU. Protein adsorption capability on polyurethane and modified-polyurethane membrane for periodontal guided tissue regeneration applications. Mater Sci Eng C. 2016;68:267–75.
148. Simion M, Scarano A, Gionso L, Piattelli A. Guided bone regeneration using resorbable and nonresorbable membranes: a comparative histologic study in humans. Int J Oral Maxillofac Implants. 1996;11(6):735–42.
149. Bunyaratavej P, Wang HL. Collagen membranes: a review. J Periodontol. 2001;72(2):215–29.
150. Patino MG, Neiders ME, Andreana S, Noble B, Cohen RE. Collagen as an implantable material in medicine and dentistry. J Oral Implantol. 2002;28(5):220–5.
151. Locci P, Calvitti M, Belcastro S, Pugliese M, Guerra M, Marinucci L, Staffolani N, Becchetti E. Phenotype expression of gingival fibroblasts cultured on membranes used in guided tissue regeneration. J Periodontol. 1997;68(9):857–63.
152. Wang HL, Miyauchi M, Takata T. Initial attachment of osteoblasts to various guided bone regeneration membranes: an in vitro study. J Periodontal Res. 2002;37(5):340–4.
153. Blumenthal NM. The use of collagen membranes to guide regeneration of new connective tissue attachment in dogs. J Periodontol. 1988;59(12):830–6.
154. Pitaru S, Tal H, Soldinger M, Grosskopf A, Noff M. Partial regeneration of periodontal tissues using collagen barriers. J Periodontol. 1988;59(6):380–6.

155. Li S-T, Chen H-C, Lee NS, Ringshia R, Yuen D. A comparative study of Zimmer BioMend® and BioMend® Extend™ membranes made at two different manufacturing facilities. Zimmer Dental White Paper; 2013. p. 1–5.
156. Shieh AT, Wang HL, O'Neal R, Glickman GN, MacNeil RL. Development and clinical evaluation of a root coverage procedure using a collagen barrier membrane. J Periodontol. 1997;68(8):770–8.
157. Wang HL, O'Neal RB, Thomas CL, Shyr Y, MacNeil RL. Evaluation of an absorbable collagen membrane in treating class II furcation defects. J Periodontol. 1994;65(11):1029–36.
158. Yukna C, Yukna R. Multi-center evaluation of bioabsorbable collagen membrane for guided tissue regeneration in human class II furcations. J Periodontol. 1996;67(7):650–7.
159. Tanner MG, Solt CW, Vuddhakanok S. An evaluation of new attachment formation using a Microfibhllar collagen barrier. J Periodontol. 1988;59(8):524–30.
160. Paul BF, Mellonig JT, Towle HJ III, Gray JL. Use of a collagen barrier to enhance healing in human periodontal furcation defects. Int J Periodont Restor Dent. 1992;12(2):123–31.
161. Kuo SM, Chang SJ, Chen TW, Kuan TC. Guided tissue regeneration for using a chitosan membrane: an experimental study in rats. J Biomed Mater Res Part A. 2006;76(2):408–15.
162. Kim HW, Song JH, Kim HE. Nanofiber generation of gelatin–hydroxyapatite biomimetics for guided tissue regeneration. Adv Funct Mater. 2005;15(12):1988–94.
163. Ling LJ, Hung SL, Lee CF, Chen YT, Wu KM. The influence of membrane exposure on the outcomes of guided tissue regeneration: clinical and microbiological aspects. J Periodontal Res. 2003;38(1):57–63.
164. Garrett S, Martin M, Egelberg J. Treatment of periodontal furcation defects Coronally positioned flaps versus dura mater membranes in class II defects. J Clin Periodontol. 1990;17(3):179–85.
165. Yukna RA. Clinical human comparison of expanded polytetrafluoroethylene barrier membrane and freeze-dried dura mater allografts for guided tissue regeneration of lost periodontal support. I. Mandibular molar class II furcations. J Periodontol. 1992;63(5):431–42.
166. Scott TA, Towle HJ, Assad DA, Nicoll BK. Comparison of bioabsorbable laminar bone membrane and non-resorbable ePTFE membrane in mandibular furcations. J Periodontol. 1997;68(7):679–86.
167. Tayebi L, Rasoulianboroujeni M, Moharamzadeh K, Almela TK, Cui Z, Ye H. 3D-printed membrane for guided tissue regeneration. Mater Sci Eng C. 2018;84:148–58.
168. Su K, Wang C. Recent advances in the use of gelatin in biomedical research. Biotechnol Lett. 2015;37(11):2139–45.
169. Shabafrooz V, Mozafari M, Köhler GA, Assefa S, Vashaee D, Tayebi L. The effect of hyaluronic acid on biofunctionality of gelatin–collagen intestine tissue engineering scaffolds. J Biomed Mater Res A. 2014;102(9):3130–9.
170. Lee AY, Han B, Lamm SD, Fierro CA, Han H-C. Effects of elastin degradation and surrounding matrix support on artery stability. Am J Phys Heart Circ Phys. 2011;302(4):H873–84.
171. Entwistle J, Hall CL, Turley EA. HA receptors: regulators of signalling to the cytoskeleton. J Cell Biochem. 1996;61(4):569–77.
172. Dehghani S, Rasoulianboroujeni M, Ghasemi H, Keshel SH, Nozarian Z, Hashemian MN, Zarei-Ghanavati M, Latifi G, Ghaffari R, Cui Z. 3D-printed membrane as an alternative to amniotic membrane for ocular surface/conjunctival defect reconstruction: an in vitro & in vivo study. Biomaterials. 2018;174:95–112.
173. Von Arx T, Cochran DL, Schenk R, Buser D. Evaluation of a prototype trilayer membrane (PTLM) for lateral ridge augmentation: an experimental study in the canine mandible. Int J Oral Maxillofac Surg. 2002;31(2):190–9.
174. Simion M, Misitano U, Gionso L, Salvato A. Treatment of dehiscences and fenestrations around dental implants using resorbable and nonresorbable membranes associated with bone autografts: a comparative clinical study. Int J Oral Maxillofac Implants. 1997;12(2):159–67.
175. Lorenzoni M, Pertl C, Keil C, Wegscheider WA. Treatment of peri-implant defects with guided bone regeneration: a comparative clinical study with various membranes and bone grafts. Int J Oral Maxillofac Implants. 1998;13(5):639–46.

176. Nair LS, Laurencin CT. Biodegradable polymers as biomaterials. Prog Polym Sci. 2007;32(8–9):762–98.
177. Haidar ZS. Bio-inspired/−functional colloidal core-shell polymeric-based nanosystems: technology promise in tissue engineering, bioimaging and nanomedicine. Polymers. 2010;2(3):323–52.
178. Donos N, Kostopoulos L, Karring T. Alveolar ridge augmentation using a resorbable copolymer membrane and autogenous bone grafts. Clin Oral Implants Res. 2002;13(2):203–13.
179. Stavropoulos F, Dahlin C, Ruskin JD, Johansson C. A comparative study of barrier membranes as graft protectors in the treatment of localized bone defects: an experimental study in a canine model. Clin Oral Implants Res. 2004;15(4):435–42.
180. Lundgren D, Mathisen T, Gottlow J. The development of a bioresorbable barrier for guided tissue regeneration. Swed Dent J. 1994;86:741–56.
181. Tengan KS. Prospective, Comparative Assessment of Alveolar Ridge Preservation Using Guidor® Easy-Graft® Classic In Atramatic Extraction Socket (Doctoral dissertation, The University of Iowa). 2017.
182. Shim J-H, Yoon M-C, Jeong C-M, Jang J, Jeong S-I, Cho D-W, Huh J-B. Efficacy of rhBMP-2 loaded PCL/PLGA/β-TCP guided bone regeneration membrane fabricated by 3D printing technology for reconstruction of calvaria defects in rabbit. Biomed Mater. 2014;9(6):065006.
183. Won J, Park C, Bae J, Ahn G, Kim C, Lim D, Cho D, Yun W, Shim J, Huh J. Evaluation of 3D printed PCL/PLGA/β-TCP versus collagen membranes for guided bone regeneration in a beagle implant model. Biomed Mater. 2016;11(5):055013.
184. Jamalpour MR, Vahdatinia F, Amirabad LM, Jamshidi S, Yadegari A, Setareh S, Moeinifard E, Tayebi L. To be published.
185. Wadhawan A, Gowda TM, Mehta DS. Gore-tex® versus resolut adapt® GTR membranes with perioglas® in periodontal regeneration. Contemp Clin Dent. 2012;3(4):406.
186. Labate GDU, Catapano G. Membranes for regenerative medicine in clinical applications. Biomed Membr (Bio) Artif Org. 2017;2:263.
187. Haghighat A, Shakeri S, Mehdikhani M, Dehnavi SS, Talebi A. Histologic, histomorphometric, and osteogenesis comparative study of a novel fabricated nanocomposite membrane versus cytoplast membrane. J Oral Maxillofac Surg. 2019;77:2027–39.
188. Li S-T, Yuen D, Martin D, Lee NS. A comparative study of a new porcine collagen membrane to Bio-Gide®. In: Science, Technology, Innovation; 2015. p. 1–5.
189. Wu S-Y, Chen Y-T, Chen C-W, Chi L-Y, Hsu N-Y, Hung S-L, Ling L-J. Comparison of clinical outcomes following guided tissue regeneration treatment with a polylactic acid barrier or a collagen membrane. Int J Periodont Restor Dent. 2010;30(2):173–9.
190. Sheikh Z, Qureshi J, Alshahrani AM, Nassar H, Ikeda Y, Glogauer M, Ganss B. Collagen based barrier membranes for periodontal guided bone regeneration applications. Odontology. 2017;105(1):1–12.
191. Seciu A-M, Gaspar A, Stefan LM, Moldovan L, Craciunescu O, Zarnescu O. Riboflavin cross-linking of collagen porous scaffolds for periodontal regeneration. Studia Universitatis Vasile Goldis Seria Stiintele Vietii (Life Sciences Series). 2016;26(2):243–9.
192. Gentile P, Chiono V, Tonda-Turo C, Ferreira AM, Ciardelli G. Polymeric membranes for guided bone regeneration. Biotechnol J. 2011;6(10):1187–97.
193. Gielkens PF, Schortinghuis J, De Jong JR, Raghoebar GM, Stegenga B, Bos RR. Vivosorb®, Bio-Gide®, and Gore-Tex® as barrier membranes in rat mandibular defects: an evaluation by microradiography and micro-CT. Clin Oral Implants Res. 2008;19(5):516–21.
194. Tabatabaei F, Tayebi L et al. unpublished data.

Printed in the United States
by Baker & Taylor Publisher Services